T0214521

Lecture Notes in Bioinformatics 11228

Subseries of Lecture Notes in Computer Science

More information about this series at http://www.springer.com/series/5381

Ronnie Alves (Ed.)

Advances in Bioinformatics and Computational Biology

11th Brazilian Symposium on Bioinformatics, BSB 2018
Niterói, Brazil, October 30 – November 1, 2018
Proceedings

Editor
Ronnie Alves (iD)
Instituto Tecnológico Vale
Belém
Brazil

ISSN 0302-9743   ISSN 1611-3349 (electronic)
Lecture Notes in Bioinformatics
ISBN 978-3-030-01721-7   ISBN 978-3-030-01722-4 (eBook)
https://doi.org/10.1007/978-3-030-01722-4

Library of Congress Control Number: 2018956561

LNCS Sublibrary: SL8 – Bioinformatics

This Springer imprint is published by the registered company Springer Nature Switzerland AG
The registered company address is: Gewerbestrasse 11, 6330 Cham, Switzerland

Preface

This volume contains the papers selected for presentation at the 11th Brazilian Symposium on Bioinformatics (BSB 2018), held from October 30 to November 1, 2018, in Niteroi, Brazil. BSB is an international conference that covers all aspects of bioinformatics and computational biology. The event was organized by the Special Interest Group in Computational Biology of the Brazilian Computer Society (SBC), which has been the organizer of BSB for the past several years. The BSB series started in 2005. In the period 2002–2004 its name was Brazilian Workshop on Bioinformatics (WOB).

As in previous editions, BSB 2018 had an international Program Committee (PC) of 31 members. After a rigorous review process by the PC, 13 papers were accepted to be orally presented at the event, and are published in this volume. All papers were reviewed by at least three independent reviewers. We believe that this volume represents a fine contribution to current research in computational biology and bioinformatics, as well as in molecular biology. In addition to the technical presentations, BSB 2018 featured keynote talks from João Meidanis (Universidade Estadual de Campinas), Peter Stadler (University of Leipzig), and David Sankoff (University of Ottawa).

BSB 2018 was made possible by the dedication and work of many people and organizations. We would like to express our sincere thanks to all PC members, as well as to the external reviewers. Their names are listed herein. We are also grateful to the local organizers and volunteers for their valuable help; the sponsors for making the event financially viable; and Springer for agreeing to publish this volume. Finally, we would like to thank all authors for their time and effort in submitting their work and the invited speakers for having accepted our invitation.

November 2018   Ronnie Alves

Organization

Conference Chairs

Daniel Cardoso de Oliveira   Universidade Federal Fluminense, Brazil
Luis Antonio Kowada   Universidade Federal Fluminense, Brazil
Ronnie Alves   Instituto Tecnológico Vale, Brazil

Local Organizing Committee

André Cunha Ribeiro	Instituto Federal de Educação Ciência e Tecnologia, Brazil
Daniel Cardoso de Oliveira	Universidade Federal Fluminense, Brazil
Helene Leitão	Universidade Federal Fluminense, Brazil
Luis Antonio Kowada	Universidade Federal Fluminense, Brazil
Luís Felipe Ignácio Cunha	Universidade Federal do Rio de Janeiro, Brazil
Simone Dantas	Universidade Federal Fluminense, Brazil
Raquel Bravo	Universidade Federal Fluminense, Brazil

Program Chair

Ronnie Alves   Instituto Tecnológico Vale, Brazil

Steering Committee

Guilherme Pimentel Telles	Universidade Estadual de Campinas, Brazil
João Carlos Setubal	Universidade de São Paulo, Brazil
Luciana Montera	Universidade Federal de Mato Grosso do Sul, Brazil
Luis Antonio Kowada	Universidade Federal Fluminense, Brazil
Maria Emilia Telles Walter	Universidade de Brasilia, Brazil
Nalvo Franco de Almeida Jr.	Universidade Federal de Mato Grosso do Sul, Brazil
Natália Florencio Martins	Empresa Brasileira de Pesquisa Agropecuária, Brazil
Ronnie Alves	Instituto Tecnológico Vale, Brazil
Sérgio Vale Aguiar Campos	Universidade Federal de Minas Gerais, Brazil
Tainá Raiol	Fundação Oswaldo Cruz, Brazil

Program Committee

Alexandre Paschoal	Universidade Federal Tecnológica do Paraná, Brazil
André C. P. L. F. de Carvalho	Universidade de São Paulo, Brazil
André Kashiwabara	Universidade Federal Tecnológica do Paraná, Brazil
Annie Chateau	Université de Montpellier, France
César Manuel Vargas Benítez	Universidade Federal Tecnológica do Paraná, Brazil

Fabrício Martins Lopes	Universidade Federal Tecnológica do Paraná, Brazil
Felipe Louza	Universidade de São Paulo, Brazil
Fernando Luís Barroso Da Silva	Universidade de São Paulo, Brazil
Guilherme Pimentel Telles	Universidade Estadual de Campinas, Brazil
Ivan G. Costa	RWTH Aachen University, Germany
Jefferson Morais	Universidade Federal do Pará, Brazil
Jens Stoye	Bielefeld University, Germany
João Carlos Setubal	Universidade de São Paulo, Brazil
Kleber Padovani de Souza	Universidade Federal do Pará, Brazil
Laurent Bréhélin	Université de Montpellier, France
Luciano Antonio Digiampietri	Universidade de São Paulo, Brazil
Luís Felipe Ignácio Cunha	Universidade Federal do Rio de Janeiro, Brazil
Luis Antonio Kowada	Universidade Federal Fluminense, Brazil
Marcilio De Souto	Université d'Orléans, France
Marcio Dorn	Universidade Federal do Rio Grande do Sul, Brazil
Maria Emilia Telles Walter	Universidade de Brasilia, Brazil
Mariana Recamonde-Mendoza	Universidade Federal do Rio Grande do Sul, Brazil
Marilia Braga	Bielefeld University, Germany
Nalvo Franco de Almeida Jr.	Universidade Federal de Mato Grosso do Sul, Brazil
Rommel Ramos	Universidade Federal do Pará, Brazil
Ronnie Alves	Instituto Tecnológico Vale, Brazil
Said Sadique Adi	Universidade Federal de Mato Grosso do Sul, Brazil
Sérgio Vale Aguiar Campos	Universidade Federal de Minas Gerais, Brazil
Sergio Lifschitzs	Pontifícia Universidade Católica do Rio de Janeiro, Brazil
Sergio Pantanos	Institut Pasteur de Montevideo, Uruguay
Tainá Raiol	Fundação Oswaldo Cruz, Brazil

Additional Reviewers

Clement Agret	CIRAD, France
Diego P. Rubert	Universidade Federal de Mato Grosso do Sul, Brazil
Eloi Araujo	Universidade Federal de Mato Grosso do Sul, Brazil
Euler Garcia	Universidade de Brasilia, Brazil
Fábio Henrique Viduani Martinez	Universidade Federal de Mato Grosso do Sul, Brazil
Fábio Vicente	Universidade Federal Tecnológica do Paraná, Brazil
Francisco Neves	Universidade de Brasilia, Brazil
Maria Beatriz Walter Costa	Universidade de Brasilia, Brazil
Pedro Feijão	Simon Fraser University, Canada
Rodrigo Hausen	Universidade Federal do ABC, Brazil
Sèverine Bérard	Université de Montpellier, France
Waldeyr Mendes	Universidade de Brasilia, Brazil

Sponsors

Sociedade Brasileira de Computação (SBC)
Universidade Federal Fluminense (UFF)
Instituto de Computação, UFF
Coordenação de Aperfeiçoamento de Pessoal de Nível Superior (CAPES)
Pró-Reitoria de Pesquisa, Pós-graduação e Inovação, UFF
Springer

Contents

Sorting λ-Permutations by λ-Operations

Guilherme Henrique Santos Miranda[1](✉)[iD],
Alexsandro Oliveira Alexandrino[1][iD], Carla Negri Lintzmayer[2][iD],
and Zanoni Dias[1][iD]

[1] Institute of Computing, University of Campinas (Unicamp), Campinas, Brazil
{guilherme.miranda,alexsandro.alexandrino}@students.ic.unicamp.br,
zanoni@ic.unicamp.br
[2] Center for Mathematics, Computation and Cognition,
Federal University of ABC (UFABC), Santo André, Brazil
carla.negri@ufabc.edu.br

Abstract. The understanding of how different two organisms are is one of the challenging tasks of modern science. A well accepted way to estimate the evolutionary distance between two organisms is estimating the rearrangement distance, which is the smallest number of rearrangements needed to transform one genome into another. If we represent genomes as permutations, we can represent one as the identity permutation and so we reduce the problem of transforming one permutation into another to the problem of sorting a permutation using the minimum number of operations. In this work, we study the problems of sorting permutations using reversals and/or transpositions, with some additional restrictions of biological relevance. Given a value λ, the problem now is how to sort a λ-permutation, which is a permutation where all elements are less than λ positions away from their correct places (regarding the identity), by applying the minimum number of operations. Each λ-operation must have size at most λ and, when applied over a λ-permutation, the result should also be a λ-permutation. We present algorithms with approximation factors of $O(\lambda^2)$, $O(\lambda)$, and $O(1)$ for the problems of Sorting λ-Permutations by λ-Reversals, by λ-Transpositions and by both operations.

Keywords: Genome rearrangements · Approximation algorithms Sorting permutations

1 Introduction

One challenge of modern science is to understand how species evolve, considering that new organisms arise from mutations that occurred in others. Using the *principle of parsimony*, the minimum number of rearrangements that transform one genome into another, called *rearrangement distance*, is considered a well accepted way to estimate the evolutionary distance between two genomes. A *genome rearrangement* is a global mutation that alters the order and/or the orientation of the genes in a genome.

R. Alves (Ed.): BSB 2018, LNBI 11228, pp. 1–13, 2018.
https://doi.org/10.1007/978-3-030-01722-4_1

Depending on the genomic information available and the problems considered, a genome can be modeled in different ways. Considering that a genome has no repeated genes, we can model it as a *permutation*, with each element representing a gene or a genomic segment shared by the genomes being compared. If the information about orientation is available, we use signed permutations to represent them. When this information is unknown, we simply use unsigned permutations. In this work, we only deal with the case where the orientation is unknown. By representing one of the genomes as the identity permutation, we reduce the problem of transforming one permutation into another to the problem of sorting a permutation with the minimum number of rearrangements, which is called sorting rearrangement distance or simply *distance*.

A *rearrangement model M* is the set of valid rearrangements used to calculate the distance. A *reversal* rearrangement inverts a segment of the genome and a *transposition* rearrangement swaps two adjacent segments of the genome. Genome rearrangements are also sometimes called operations.

The problems of Sorting Permutations by Reversals and Sorting Permutations by Transpositions are NP-Hard [2,3]. The best-known results for both problems are approximation algorithms with factor 1.375 [1,6]. Walter *et al.* [13] considered a variation in which the rearrangement model contains both reversals and transpositions, which is called Sorting Permutations by Reversals and Transpositions, and they presented a 3-approximation algorithm for it. However, the best-known approximation factor for this problem is $2k$ [11], where k is the approximation factor of an algorithm used for cycle decomposition of the breakpoint graph [5]. Given the best-known value for k, this algorithm has an approximation factor of $2.8386 + \epsilon$, where $\epsilon > 0$. The complexity of this problem is still unknown.

Many variants of the sorting rearrangement distance emerged from the assumption that rearrangement operations which affect large portions of a genome are less likely to occur [9]. Most of these variants add a constraint that limits the size of a valid operation [7,8,12]. Considering a size-limit of 2, the problems of Sorting Permutations by Reversals and/or Transpositions are solvable in polynomial time [7]. Considering a size-limit of 3, the best-known approximation factors for Sorting Permutations by Reversals, by Transpositions, and by Reversals and Transpositions are 2 [7], 5/4 [8], and 2 [12], respectively.

The problem of Sorting Permutations by λ-operations is a generalization of the size-limited variants in which a rearrangement operation is valid if its size is less than or equal to λ. Miranda *et al.* [10] presented $O(\lambda^2)$-approximation algorithms for reversals, transpositions, and both operations. Using size-limited operations makes more sense when one knows that the elements are not so far away from their original positions, so we introduce the study of Sorting λ-Permutations by λ-Operations. A permutation π is a λ-*permutation* if all elements of π are at a distance less than λ from their correct positions considering the identity permutation. We will consider the problems of sorting unsigned λ-permutations by λ-reversals, by λ-transpositions, and by λ-reversals and λ-transpositions.

Next sections are organized as follows. In Sect. 2, we present all the concepts used in this paper. In Sect. 3, we present $O(\lambda^2)$-approximation algorithms for the problems studied. In Sect. 4, we present algorithms with approximation factors of $O(\lambda)$ and $O(1)$. In Sect. 5, we show experimental results which compare the algorithms presented in Sects. 3 and 4. We conclude the paper in Sect. 6.

2 Definitions

We denote a permutation by $\pi = (\pi_1 \pi_2 \ldots \pi_n)$, where $\pi_i \in \{1, 2, \ldots, n\}$ and $\pi_i \neq \pi_j$, for all $1 \leq i < j \leq n$. We assume that there are two extra elements $\pi_0 = 0$ and $\pi_{n+1} = n + 1$ in π, but, for convenience, they are omitted from the permutation's representation. Given an integer λ as input, we say that π is a λ-*permutation* if we have $|\pi_i - i| < \lambda$ for all $1 \leq i \leq n$.

We denote the inverse permutation of π as π^{-1}. This permutation is such that $\pi^{-1}_{\pi_i} = i$ for all $1 \leq i \leq n$. Note that element π_i^{-1} indicates the position of element i in π. For example, given $\pi = (4\ 6\ 3\ 5\ 2\ 1)$, then $\pi^{-1} = (6\ 5\ 3\ 1\ 4\ 2)$.

An operation of *reversal* is denoted by $\rho(i, j)$, with $1 \leq i < j \leq n$, and when applied on a permutation $\pi = (\pi_1 \pi_2 \ldots \pi_n)$, the result is permutation $\pi \cdot \rho(i, j)$ $= (\pi_1 \pi_2 \ldots \pi_{i-1} \underline{\pi_j \pi_{j-1} \ldots \pi_{i+1} \pi_i} \pi_{j+1} \ldots \pi_{n-1} \pi_n)$. The size of a reversal is given by $j - i + 1$. For example, given $\pi = (1\ 4\ 3\ 2\ 5)$, then $\pi \cdot \rho(2, 4) = (1\ 2\ 3\ 4\ 5)$ and the size of such operation is $4 - 2 + 1 = 3$. We say that $\rho(i, j)$ is a λ-*reversal* if we have $j - i + 1 \leq \lambda$.

An operation of *transposition* is denoted by $\tau(i, j, k)$, with $1 \leq i < j < k \leq n + 1$, and when applied on a permutation $\pi = (\pi_1 \pi_2 \ldots \pi_n)$, the result is permutation $\pi \cdot \tau(i, j, k) = (\pi_1 \pi_2 \ldots \pi_{i-1} \underline{\pi_j \ldots \pi_{k-1} \pi_i \ldots \pi_{j-1}} \pi_k \ldots \pi_n)$. The size of a transposition is given by $k - i$. For example, given $\pi = (4\ 5\ 6\ 1\ 2\ 3)$, then $\pi \cdot \tau(1, 4, 7) = (1\ 2\ 3\ 4\ 5\ 6)$ and the size of such operation is $7 - 1 = 6$. We say that $\tau(i, j, k)$ is a λ-*transposition* if we have $k - i \leq \lambda$.

The goal of these problems is to transform a λ-permutation π into the *identity permutation* $\iota = (1\ 2\ \ldots\ n)$ by applying the minimum number of λ-operations, which defines the *sorting distance*, such that each permutation generated during the process is also a λ-permutation.

We denote by $d_r^\lambda(\pi)$, $d_t^\lambda(\pi)$, and $d_{rt}^\lambda(\pi)$, the sorting distance when we have only λ-reversals, only λ-transpositions, and when we have both operations, respectively.

3 Inversions-Based Approximation Algorithms

In this section we present approximation algorithms based on the concept of inversions for the problems we are addressing.

An *inversion* is defined as a pair of elements (π_i, π_j) such that $i < j$ and $\pi_i > \pi_j$. The number of inversions in π is denoted by $\mathrm{Inv}(\pi)$.

Lemma 1. *For all λ-permutations $\pi \neq \iota$ and all $\lambda \geq 2$, we have $d_r^\lambda(\pi) \geq \frac{\mathrm{Inv}(\pi)}{\lambda(\lambda-1)/2}$, $d_t^\lambda(\pi) \geq \frac{\mathrm{Inv}(\pi)}{\lambda(\lambda-1)/2}$, and $d_{rt}^\lambda(\pi) \geq \frac{\mathrm{Inv}(\pi)}{\lambda(\lambda-1)/2}$.*

Proof. This follows immediately from the observation that a λ-reversal and a λ-transposition can remove at most $\frac{\lambda(\lambda-1)}{2}$ inversions.   □

Next lemma shows that it is always possible to remove at least one inversion from a λ-permutation by applying one λ-operation which results in a λ-permutation.

Lemma 2. *Let π be a λ-permutation. It is always possible to obtain a λ-permutation with $\mathrm{Inv}(\pi) - 1$ inversions by applying one λ-reversal or one λ-transposition.*

Proof. Let $\pi_j = i$ be the smallest element out-of-place ($\pi_i \neq i$) in π. Initially, note we have inversion (π_{j-1}, π_j), since $\pi_{j-1} > \pi_j$ and $j - 1 < j$. Let σ be a λ-operation that swaps elements π_{j-1} and π_j, and let $\pi' = \pi \cdot \sigma$. It is easy to see that $\mathrm{Inv}(\pi') = \mathrm{Inv}(\pi) - 1$, since such inversion was removed, and note that there always exists a λ-reversal or a λ-transposition equivalent to σ, because the elements are adjacent and so both only swap two elements.

Observe that, in π', element π_j is closer to its correct position, since it was moved to the left. Hence, we follow by showing that π' is a λ-permutation by considering two cases according to the values of $\pi_{j-1} = \pi'_j$.

If $\pi_{j-1} \geq j$, then π_{j-1} is also closer to its correct position, in π'. Otherwise, $\pi_{j-1} < j$. Thus, element π_{j-1} will be, in π', one position away from its correct position. Then, note that $|j - \pi_j| + 1 = (j - i) + 1 \leq \lambda$, because π is a λ-permutation. Also observe that we have $|j - \pi'_j| = j - \pi'_j$ because $\pi'_j = \pi_{j-1} < j$, and $j - \pi'_j < j - i$ because $\pi'_j > i$, and, so, $j - i \leq \lambda - 1$. Therefore, π' is a λ-permutation and the result follows.   □

A generic greedy approximation algorithm for the three problems we are addressing is presented in next theorem. It receives an integer $\lambda \geq 2$ and a λ-permutation $\pi \neq \iota$ as inputs. It is greedy because it always tries to decrease the largest amount of inversions in π. Since the only permutation with no inversions is the identity, it will, eventually, sort π.

Theorem 3. *There exist $O(\lambda^2)$-approximation algorithms for the problems of Sorting λ-Permutations by λ-Reversals, λ-Transpositions, and by λ-Reversals and λ-Transpositions.*

Proof. Let $\lambda \geq 2$ be an integer and let $\pi \neq \iota$ be a λ-permutation. Consider an algorithm which chooses the λ-operation σ such that $\pi \cdot \sigma$ is a λ-permutation and $\mathrm{Inv}(\pi \cdot \sigma)$ is as small as possible and then it applies such operation over π. The algorithm repeats the same process in the resulting permutation until it reaches the identity permutation.

In the worst case, we always have one λ-operation reducing the number of inversions by one unit, as shown in Lemma 2. Therefore, the number of operations of such greedy algorithm is at most $\mathrm{Inv}(\pi)$, and the approximation factor follows immediately from Lemma 1.   □

Note that the distance is $O(n^2)$ because any permutation can be sorted with $O(n^2)$ λ-reversals or λ-transpositions. For Sorting by λ-Reversals, at each step the algorithm considers $O(\lambda^2)$ possible reversals that can be chosen. Since the variation in the number of inversions caused by an operation can be calculated in $O(\lambda\sqrt{\log \lambda})$ time [4], the algorithm has total time complexity $O(n^2\lambda^3\sqrt{\log \lambda})$. Using the same analysis, we conclude that the algorithms involving transpositions have total time complexity $O(n^2\lambda^4\sqrt{\log \lambda})$.

4   Breakpoints-Based Approximation Algorithms

In this section we present approximation algorithms based on the concept of breakpoints for the three problems we are addressing.

A *breakpoint* is defined as a pair of elements (π_i, π_{i+1}) such that $\pi_{i+1} - \pi_i \neq 1$ (resp. $|\pi_{i+1} - \pi_i| \neq 1$), for all $0 \leq i \leq n$, in the problems of Sorting λ-Permutations by λ-Transpositions (resp. Sorting λ-Permutations by λ-Reversals and Sorting λ-Permutations by λ-Reversals and λ-Transpositions). The number of breakpoints in π is denoted by $b(\pi)$.

Lemma 4. *For all λ-permutations $\pi \neq \iota$ and all $\lambda \geq 2$ we have $d_r^\lambda(\pi) \geq \frac{b(\pi)}{2}$, $d_t^\lambda(\pi) \geq \frac{b(\pi)}{3}$, and $d_{rt}^\lambda(\pi) \geq \frac{b(\pi)}{3}$.*

Proof. This follows immediately from the observation that a λ-reversal and a λ-transposition can remove at most 2 and 3 breakpoints, respectively. □

A maximal subsequence $(\pi_i\,\pi_{i+1} \ldots \pi_j)$ without any breakpoints (π_k, π_{k+1}), for all $i \leq k < j$, is called a *strip*. If the strip's elements are in ascending (resp. descending) order, then we call it an *increasing* (resp. *decreasing*) strip. Strips containing only one element are considered to be increasing. For example, considering sorting by both operations and $\pi = (6\ 4\ 5\ 3\ 2\ 1)$, we have two increasing strips (6) and (4 5), and a decreasing strip (3 2 1). Note, however, that segment (3 2 1) is not a decreasing strip for the problem of Sorting by Transpositions. This is because, direct from the definition of strips and breakpoints, there are no decreasing strips when this problem is considered. The number of elements in a strip S of a λ-permutation π is denoted by $|S|$.

Lemma 5. *Let π be a λ-permutation and let $\pi_j = i$ be the smallest element out-of-place in π. The strip S that contains π_j is such that $|S| \leq \lambda - 1$.*

Proof. First, suppose we have $S = (\pi_j \ldots \pi_k)$ as an increasing strip. Then, let $R = (\pi_i \ldots \pi_{j-1})$ be the segment of elements to the left of S. Note that any element in R is greater than any element in S and, so, $\pi_i \geq k + 1$, since $\pi_i > \pi_k$. By contradiction, suppose $|S| \geq \lambda$. Then, we have $i \leq k - \lambda + 1$. Since $i \leq (k+1) - \lambda \leq k + 1 \leq \pi_i$, we also have that $i \leq \pi_i$, and so $|\pi_i - i| = \pi_i - i \geq k + 1 - (k + 1 - \lambda) = \lambda > \lambda - 1$, which is a contradiction to the definition of λ-permutations. When we have π_j in a decreasing strip, the proof follows by symmetry. □

In next lemma, we suppose that the smallest element out-of-place is in an increasing strip of a λ-permutation $\pi \neq \iota$ and we show how to reduce the number of breakpoints of π by moving this strip to its correct position, but without considering λ-operations. It is auxiliar to Lemmas 7 and 8, which show how to do this by applying a sequence of λ-transpositions. Lemma 9 shows how to do this when the smallest element out-of-place is in a decreasing strip.

Lemma 6. *Let π be a λ-permutation. Let $\pi_j = i$ be the smallest element out-of-place in π. Suppose that π_j is in an increasing strip $S = (\pi_j \ldots \pi_k)$. Then $b(\pi \cdot \tau(i, j, k+1)) \leq b(\pi) - 1$ and $(k + 1 - i) \leq 2(\lambda - 1)$.*

Proof. Let $R = (\pi_i \ldots \pi_{j-1})$ be the segment of elements in π that will be transposed with S. Observe that any element in R is greater than any element in S, so $\pi \cdot \tau(i, j, k+1)$ is a λ-permutation, once greater elements are moved to the right and smaller elements to the left. Also observe that in π we have the three breakpoints (π_{i-1}, π_i), (π_j, π_{j+1}), and (π_{k-1}, π_k), where the first one is because $\pi_{i-1} = \pi_j - 1 = i - 1$ and $\pi_i > i = \pi_j$ and the second and third ones are because the strip's start and strip's end are at positions j and k, respectively. Transposition $\tau(i, j, k+1)$ moves the elements of S to their correct positions by transposing them with elements of R, thus removing at least breakpoint (π_{i-1}, π_i). Since a transposition can add at most three breakpoints, but we already had all of them and we removed at least (π_{i-1}, π_i), we have $b(\pi \cdot \tau(i, j, k+1)) \leq b(\pi) - 1$.

By Lemma 5, we have $|S| \leq \lambda - 1$, thus $k + 1 - j \leq \lambda - 1$. Since π is a λ-permutation, we have $|\pi_j - j| \leq \lambda - 1$, and, by construction, $\pi_j = i$, thus $|i - j| + 1 = j - i + 1 \leq \lambda - 1$. Therefore, $k + 1 - i \leq 2(\lambda - 1)$.   □

Lemma 7. *Let π be a λ-permutation. Let $\pi_j = i$ be the smallest element out-of-place in π. Suppose that π only has increasing strips and that π_j is in a strip $S = (\pi_j \ldots \pi_k)$. It is always possible to obtain a λ-permutation $\pi \cdot \tau(i, j, k+1)$ with at most $b(\pi) - 1$ breakpoints by applying at most $5 + \lceil (\lambda - 1)/2 \rceil$ λ-reversals such that all intermediary permutations are λ-permutations.*

Proof. Let $R = (\pi_i \ldots \pi_{j-1})$ be the segment that will be moved to the right in $\tau(i, j, k+1)$. Note that $|S| \leq \lambda - 1$, by Lemma 5, and $|R| \leq \lambda - 1$, because π is a λ-permutation.

The idea is to move elements from S to their correct positions by applying at most two sequences of pairs of λ-reversals, where each one puts at most $\lfloor \lambda/2 \rfloor$ elements of S in their correct positions at a time.

In the first sequence of λ-reversals, there are two possibilities. If $|S| \leq \lfloor \lambda/2 \rfloor$, then the first operation of each pair reverts $|S|$ elements contained in both S and R. If $|S| > \lfloor \lambda/2 \rfloor$, then it reverts $\lfloor \lambda/2 \rfloor$ elements contained in both S and R. In any case, the second operation of each pair reverts back the elements of R affected by the first one, in order to leave π with only increasing strips again (except for the elements of S which were affected by the first operation).

After the sequence is applied, we have at most $\lfloor \lambda/2 \rfloor$ elements of S from positions i to $i + \min(\lfloor \lambda/2 \rfloor, |S|)$, and, maybe, they are in a decreasing strip. If this is the case, then one more λ-reversal has to be applied to put these elements in their correct places, by reversing such decreasing strip.

The second sequence of λ-reversals puts the at most $\lfloor \lambda/2 \rfloor$ remaining elements of S in their correct positions, following the same idea, and, also, maybe one extra λ-reversal will be necessary after it is applied. Note that, if there are no remaining elements (in case of $|S| \leq \lfloor \lambda/2 \rfloor$), this sequence is not necessary.

The largest amount of operations needed in the process described above happens when we have exactly $\lfloor \lambda/2 \rfloor + 1$ elements in S. In order to put the first $\lfloor \lambda/2 \rfloor$ elements of S in their correct positions, in this case, we have to apply at most $\lceil (\lambda - 1)/(\lfloor (\lambda/2) \rfloor + 1) \rceil$ pairs of λ-reversals and, maybe, one extra at the end. Since each pair of operations puts $\lfloor \lambda/2 \rfloor$ elements exactly (except, maybe, by the last pair) $\lfloor \lambda/2 \rfloor + 1$ positions to the left, the number of operations needed is $4 + 1 = 5$. Then, to move the remaining element of S to its correct position, all the λ-reversals of the second sequence will have size 2 (note that, in this case, we do not need the second operation of each pair), which means such element will be moved only 2 positions to the left per operation, giving an extra amount of $\lceil (\lambda - 1)/2 \rceil$ λ-reversals. Therefore, the number of λ-reversals to move S to its correct position is at most $5 + \lceil (\lambda - 1)/2 \rceil$ λ-reversals.

Now we have to show that after each operation is applied, we have a λ-permutation as result and, after the last operation is applied, we have $\pi \cdot \tau(i, j, k+1)$. Observe that any element in R is greater than any element in S. Then, once the first operation of each pair moves elements of R to the right and elements of S to the left, all elements affected will be closer to their correct positions, resulting in a λ-permutation. The second operation of each pair reverts elements of R to ascending order again, so it also results in a λ-permutation. After both sequences of λ-reversals are applied, all elements of S are at positions from i to $i + k - j$ and all elements of R are at positions from $i + k - 1$ to k, resulting in $\pi \cdot \tau(i, j, k+1)$, which is a λ-permutation with at least one less breakpoint than π, as showed in Lemma 6. $\square$

Lemma 8. *Let π be a λ-permutation. Let $\pi_j = i$ be the smallest element out-of-place in π. Suppose that π_j is in an increasing strip $S = (\pi_j \ldots \pi_k)$. It is always possible to obtain a λ-permutation $\pi \cdot \tau(i, j, k+1)$ with at most $b(\pi) - 1$ reversal breakpoints by applying at most 4 λ-transpositions such that all intermediary permutations are λ-permutations.*

Proof. Let $R = (\pi_i \ldots \pi_{j-1})$ be the segment that will be moved to the right in $\tau(i, j, k+1)$. Note that $|S| \leq \lambda - 1$, by Lemma 5, and $|R| \leq \lambda - 1$, because π is a λ-permutation.

The idea is to apply a sequence with at most four λ-transpositions that divide both segments $R = (\pi_i \ldots \pi_{j-1})$ and $S = (\pi_j \ldots \pi_k)$ into at most two parts each, where each part has at most $\lfloor \lambda/2 \rfloor$ elements, and then exchange each part of S at most twice (and at least once), with the (possible) two parts of R. If we had exactly $\lambda - 1$ elements in each of S and R, such sequence would be $\tau(i + \lfloor \lambda/2 \rfloor, j, j + \lfloor \lambda/2 \rfloor), \tau(i, i + \lfloor \lambda/2 \rfloor, j), \tau(j, j + \lfloor \lambda/2 \rfloor, k+1), \tau(i + \lfloor \lambda/2 \rfloor, j, j + \lfloor \lambda/2 \rfloor)$.

Now we have to show that after each of the at most four operations is applied, we have a λ-permutation as result. Observe that any element in R is greater than any element in S. Since each λ-transposition puts elements of S closer to their correct positions by transposing them with greater elements of R, we have a λ-permutation after each λ-operation applied. After all λ-transpositions are applied, the elements of S are at positions from i to $i + k - j$ and the elements of R are at positions from $i + k - j + 1$ to k, resulting in $\pi \cdot \tau(i, j, k + 1)$, which is a λ-permutation with at least $b(\pi) - 1$ breakpoints, as showed in Lemma 6. $\square$

Lemma 9. *Let $\pi_k = i$ be the smallest element out-of-place in a λ-permutation π. Suppose that π_k is in a decreasing strip $S = (\pi_j \ldots \pi_k)$. It is always possible to obtain a λ-permutation with at most $b(\pi) - 1$ reversal breakpoints by applying at most one λ-transposition and one λ-reversal.*

Proof. When $j = i$, one reversal $\rho(j, k)$ put elements of S in their correct positions. Since $|S| = k - j + 1 \leq \lambda - 1$ by Lemma 5, we have $\rho(j, k)$ is a λ-reversal and, since such operation just reverts a decreasing strip, we also have $\pi \cdot \rho(j, k)$ as a λ-permutation.

Now assume $j > i$. Note that, in this case, we have the three breakpoints (π_{i-1}, π_i), (π_j, π_{j+1}), and (π_{k-1}, π_k), where the first one is because $\pi_{i-1} = \pi_k - 1 = i - 1$ and $\pi_i > i = \pi_k$ and the second and third ones are because the strip's start and strip's end are at positions k and j, respectively. Thus, we can apply the λ-transposition $\tau(i, j, k + 1)$ followed by the λ-reversal $\rho(i, i + (k - j))$ and then we get $b(\pi \cdot \tau(i, j, k + 1) \cdot \rho(i, i + (k - j))) \leq b(\pi) - 1$, once a λ-transposition can add at most three breakpoints but we already had (π_{i-1}, π_i), (π_{j-1}, π_j), and (π_k, π_{k+1}), and the second λ-reversal can add at most two breakpoints but we already had (π_{i-1}, π_j) and (π_k, π_i) and we removed the first one, since all elements of S will be in their correct positions in $\pi \cdot \tau(i, j, k + 1) \cdot \rho(i, i + (k - j))$.

Now we have to show that after each operation is applied, we have a λ-permutation as result. Let $R = (\pi_i \ldots \pi_{j-1})$ be the segment of elements that should be moved in order to put S in its correct position. Observe that any element in R is greater than any element in S. The first operation, a λ-transposition, transposes S only with greater elements and thus the result is a λ-permutation. The second operation, a λ-reversal, just reverts a decreasing strip to put the elements of S in their correct positions, thus it also results in a λ-permutation. Hence, we have as result a λ-permutation with at least one less breakpoint. $\square$

Next theorems describe approximation algorithms for the problems we are addressing. Lemma 10 is auxiliar to Theorem 11. The algorithms receive an integer $\lambda \geq 2$ and a λ-permutation $\pi \neq \iota$ as input. The goal is to decrease at least one unit on the number of breakpoints in π by moving elements to their correct positions (applying Lemmas 8 and 9). Since the only permutation with no breakpoints is the identity, they will, eventually, sort π.

Lemma 10. *Let π be a λ-permutation. Let $S = (j \ldots i)$ be a decreasing strip in π (thus $i < j$). Let $\pi' = \pi \cdot \rho(\pi_j^{-1}, \pi_i^{-1})$ be the resulting permutation after reverting S in π. Then, π' is a λ-permutation.*

Proof. First note that element π_j is to the right of element π_i. We show that the lemma follows by considering four cases, according to the positions of elements i and j with relation to the elements π_i and π_j.

Case (i), $i < j < \pi_j^{-1} < \pi_i^{-1}$: note that both π_i and π_j are to the left of S. Then, after reverting S, element i is closer to its correct position, while element j is moved away from its correct position. Despite this, the distance between π_j and j in π' is smaller than the distance between π_i and i in π, and so if π is a λ-permutation, π' is also a λ-permutation.

Case (ii), $i < \pi_i^{-1} \le j < \pi_i^{-1}$: note that π_i is to the left of S and π_j is in S. Then, after reverting S, the element i is closer to its correct position and the distance of j to its correct position will still be less than λ, since the size of S is at most λ, as Lemma 5 shows.

Case (iii), $\pi_j^{-1} \le i < \pi_i^{-1} \le j$: similar to (ii).

Case (iv), $\pi_j^{-1} < \pi_i^{-1} \le i < j$: similar to (i).   □

Theorem 11. *The problem of Sorting λ-Permutations by λ-Reversals has a* $O(\lambda)$-*approximation algorithm.*

Proof. Let $\lambda \ge 2$ be an integer and $\pi \ne \iota$ be a λ-permutation. Consider an algorithm which first applies one λ-reversal over each decreasing strip of π in order to get a λ-permutation with only increasing strips. By Lemma 10, we guarantee that all intermediary permutations generated by these λ-reversals are λ-permutations.

Then, the algorithm will repeatedly take the smallest element out-of-place and move the increasing strip that contains it to its correct position, obtaining a λ-permutation with at least one less breakpoint, until it reaches the identity permutation.

As shown in Lemma 7, at most $5 + \lceil (\lambda - 1)/2 \rceil$ λ-reversals are needed to move each strip to its correct position. Since, maybe, one extra λ-reversal could have been applied in the beginning of the algorithm to transform such strip into an increasing one, we have that at most $6 + \lceil (\lambda - 1)/2 \rceil$ λ-reversals can be applied to remove at least one breakpoint. Therefore, the number of operations of our algorithm is at most $(6 + \lceil (\lambda - 1)/2 \rceil)b(\pi) \le O(\lambda)d_r^\lambda(\pi)$, where the inequality follows from Lemma 4.   □

Theorem 12. *The problem of Sorting λ-Permutations by λ-Transpositions has a 12-approximation algorithm.*

Proof. Let $\lambda \ge 2$ be an integer and let $\pi \ne \iota$ be a λ-permutation. The algorithm will repeatedly take the smallest element out-of-place and move the increasing strip that contains such element to its correct position, obtaining a λ-permutation with at least one less breakpoint, until it reaches the identity permutation.

As shown in Lemma 8, at most 4 λ-transpositions are needed to move each strip to its correct position. Then, in the worst case, we remove 1 breakpoint every 4 λ-transpositions applied. With this and Lemma 4, the number of operations of our algorithm is at most $4b(\pi) \le 12d_t^\lambda(\pi)$.   □

Theorem 13. *The problem of Sorting λ-Permutations by λ-Reversals and λ-Transpositions has a 12-approximation algorithm.*

Proof. Let $\lambda \geq 2$ be an integer and let $\pi \neq \iota$ be a λ-permutation. Let $\pi_j = i$ be the smallest element out-of-place in π. We have two cases to consider: when the strip which contains π_j is decreasing or not. In both cases, we can at least remove breakpoint (π_{i-1}, π_i) from π without adding other ones by applying at most 4 λ-transpositions (if the strip is increasing) or at most 2 λ-operations (if the strip is decreasing), as showed in Lemmas 8 and 9, respectively.

   Then, considering both cases described, the algorithm will repeatedly take the smallest element out-of-place and move the strip that contains it to its correct position, decreasing at least one breakpoint at a time, until it reaches the identity permutation.

   Note that, in the worst case, we remove 1 breakpoint every 4 λ-transpositions and so the result is analogous to Theorem 12.   □

   Since $b(n) \leq n + 1$, the algorithms move the strip containing the smallest element out of place at most $O(n)$ times. At each step, the algorithms spend $O(n)$ time to find the strip to move and they spend $O(\lambda)$ ($O(1)$-approximation algorithms) or $O(\lambda^2)$ ($O(\lambda)$-approximation algorithms) time to apply the operations to move such strip. So, the time complexity for the $O(1)$-approximation algorithms and the $O(\lambda)$-approximation algorithm are $O(n(n+\lambda))$ and $O(n(n+\lambda^2))$, respectively.

5   Experimental Results

We have implemented the inversions-based and the breakpoints-based approximation algorithms in order to analyze how they work in a practical perspective. We performed experiments considering a total of 1000 random λ-permutations, with size equal to 100 and values of $\lambda = 5, 10, 15, \ldots, 100$, as input for the algorithms. Then, we compared the results according to the average and maximum approximation factors obtained for all permutations. For each permutation, we considered the maximum value of lower bound between the ones shown in Lemmas 1 and 4.

   We show the results in Fig. 1. We observed that the maximum approximation factors were 5.38 and 6.76 for λ-reversals, 2.91 and 3.15 for λ-transpositions, and 3.00 and 3.18 for when both operations are allowed, considering the breakpoints-based and the inversions-based algorithm, respectively. We also noticed, in our tests, the average approximation factor of the inversions-based algorithm and the breakpoints-based algorithm were similar, even with the relevant difference among their theoretical approximation factors.

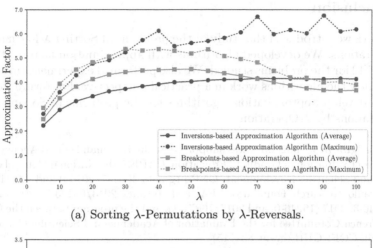

(a) Sorting λ-Permutations by λ-Reversals.

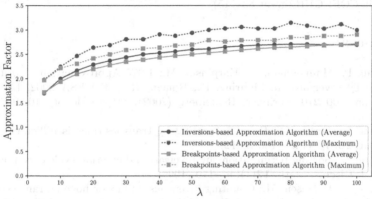

(b) Sorting λ-Permutations by λ-Transpositions.

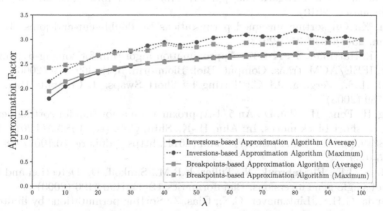

(c) Sorting λ-Permutations by λ-Reversals and λ-Transpositions.

Fig. 1. Average and maximum approximation factors of the algorithms for Sorting λ-Permutations by λ-Operations, with λ-permutations of size 100.

6   Conclusion

In this work we introduced the study of the problems of Sorting λ-Permutations by λ-Operations. We developed algorithms with approximation factors of $O(\lambda^2)$, $O(\lambda)$, and 12 for the problems studied. We also performed experiments in order to compare how the algorithms work in a practical perspective. For future work, we intend to develop approximation algorithms for the problems of Sorting Signed λ-Permutations by λ-Operations.

Acknowledgments. This work was supported by the Brazilian Federal Agency for the Support and Evaluation of Graduate Education, CAPES, the National Counsel of Technological and Scientific Development, CNPq (grants 400487/2016-0 and 425340/2016-3), São Paulo Research Foundation, FAPESP (grants 2013/08293-7, 2015/11937-9, 2017/12646-3, 2017/16246-0, and 2017/16871-1), and the program between the CAPES and the French Committee for the Evaluation of Academic and Scientific Cooperation with Brazil, COFECUB (grant 831/15).

References

1. Berman, P., Hannenhalli, S., Karpinski, M.: 1.375-Approximation algorithm for sorting by reversals. In: Möhring, R., Raman, R. (eds.) ESA 2002. LNCS, vol. 2461, pp. 200–210. Springer, Heidelberg (2002). https://doi.org/10.1007/3-540-45749-6_21
2. Bulteau, L., Fertin, G., Rusu, I.: Sorting by transpositions is difficult. SIAM J. Comput. **26**(3), 1148–1180 (2012)
3. Caprara, A.: Sorting permutations by reversals and eulerian cycle decompositions. SIAM J. Discret. Math. **12**(1), 91–110 (1999)
4. Chan, T.M., Pătraşcu, M.: Counting inversions, offline orthogonal range counting, and related problems. In: Proceedings of the Twenty-first Annual ACM-SIAM Symposium on Discrete Algorithms, pp. 161–173. Society for Industrial and Applied Mathematics (2010)
5. Chen, X.: On sorting unsigned permutations by double-cut-and-joins. J. Comb. Optim. **25**(3), 339–351 (2013)
6. Elias, I., Hartman, T.: A 1.375-Approximation algorithm for sorting by transpositions. IEEE/ACM Trans. Comput. Biol. Bioinform. **3**(4), 369–379 (2006)
7. Heath, L.S., Vergara, J.P.C.: Sorting by Short Swaps. J. Comput. Biol. **10**(5), 775–789 (2003)
8. Jiang, H., Feng, H., Zhu, D.: An 5/4-Approximation algorithm for sorting permutations by short block moves. In: Ahn, H.-K., Shin, C.-S. (eds.) ISAAC 2014. LNCS, vol. 8889, pp. 491–503. Springer, Cham (2014). https://doi.org/10.1007/978-3-319-13075-0_39
9. Lefebvre, J.F., El-Mabrouk, N., Tillier, E.R.M., Sankoff, D.: Detection and validation of single gene inversions. Bioinformatics **19**(1), i190–i196 (2003)
10. Miranda, G.H.S., Lintzmayer, C.N., Dias, Z.: Sorting permutations by limited-size operations. In: Jansson, J., Martín-Vide, C., Vega-Rodríguez, M.A. (eds.) AlCoB 2018. LNCS, vol. 10849, pp. 76–87. Springer, Cham (2018). https://doi.org/10.1007/978-3-319-91938-6_7
11. Rahman, A., Shatabda, S., Hasan, M.: An approximation algorithm for sorting by reversals and transpositions. J. Discret. Algorithms **6**(3), 449–457 (2008)

12. Vergara, J.P.C.: Sorting by Bounded Permutations. Ph.D. thesis, Virginia Poly-technic Institute and State University (1998)
13. Walter, M.E.M.T., Dias, Z., Meidanis, J.: Reversal and transposition distance of linear chromosomes. In: Proceedings of the 5th International Symposium on String Processing and Information Retrieval (SPIRE 1998), pp. 96–102. IEEE Computer Society (1998)

Super Short Reversals on Both Gene Order and Intergenic Sizes

Andre Rodrigues Oliveira[1]([⊠]) [ID], Géraldine Jean[2], Guillaume Fertin[2] [ID],
Ulisses Dias[3] [ID], and Zanoni Dias[1] [ID]

[1] Institute of Computing, University of Campinas, Campinas, Brazil
{andrero,zanoni}@ic.unicamp.br
[2] Laboratoire des Sciences du Numérique de Nantes, UMR CNRS 6004,
University of Nantes, Nantes, France
{geraldine.jean,guillaume.fertin}@univ-nantes.fr
[3] School of Technology, University of Campinas, Limeira, Brazil
ulisses@ft.unicamp.br

Abstract. The evolutionary distance between two genomes can be estimated by computing the minimum length sequence of operations, called *genome rearrangements*, that transform one genome into another. Usually, a genome is modeled as an ordered sequence of (possibly signed) genes, and almost all the studies that have been undertaken in the genome rearrangement literature consist in shaping biological scenarios into mathematical models: for instance, allowing different genome rearrangements operations at the same time, adding constraints to these rearrangements (e.g., each rearrangement can affect at most a given number k of genes), considering that a rearrangement implies a cost depending on its length rather than a unit cost, etc. However, most of the works in the field have overlooked some important features inside genomes, such as the presence of sequences of nucleotides between genes, called *intergenic regions*. In this work, we investigate the problem of computing the distance between two genomes, taking into account both gene order and intergenic sizes; the genome rearrangement operation we consider here is a constrained type of reversals, called *super short reversals*, which affect up to two (consecutive) genes. We propose here three algorithms to solve the problem: a 3-approximation algorithm that applies to any instance, and two additional algorithms that apply only on specific types of genomes with respect to their gene order: the first one is an exact algorithm, while the second is a 2-approximation algorithm.

Keywords: Genome rearrangements · Intergenic regions
Super short reversals · Approximation algorithm

1 Introduction

Given two genomes $\mathcal{G}_1$ and $\mathcal{G}_2$, one way to estimate their evolutionary distance is to compute the minimum possible number of large scale events, called *genome*

© Springer Nature Switzerland AG 2018
R. Alves (Ed.): BSB 2018, LNBI 11228, pp. 14–25, 2018.
https://doi.org/10.1007/978-3-030-01722-4_2

rearrangements, that are needed to go from G_1 to G_2. The minimality requirement is dictated by the commonly accepted parsimony principle, while the allowed genome rearrangements depend on the model, i.e. on the classes of events that supposedly happen during evolution.

However, before one performs this task, it is necessary to model the input genomes. Almost all previous works have defined genomes as ordered sequences of elements, which are *genes*. Variants within this setting can occur: for instance, depending on the model, genes may be signed or unsigned, the sign of a gene representing the DNA strand it lies on. Besides, each gene may appear either once or several times in a genome: in the latter case, genomes are modeled as strings, while in the former case they are modeled as *permutations*.

Concerning genome rearrangements, the most commonly studied is *reversal*, which consists in taking a continuous sequence in the genome, reversing it, and putting it back at the same location (see e.g. [10] for one of the first studies of the problem). A more recent and general type of genome rearrangement is the DCJ (for Double-Cut and Join) [14]. One can also alternately define the rearrangement events in order to reflect specific biological scenarios. For example, in populations where the number of rearrangement events that affect a very large portion of the genes is known to be rare, we can restrict events to be applied over no more than k genes at the same time, for some predetermined value of k [5,8,9].

Since the mid-nineties, a very large amount of work has been done concerning algorithmic issues of computing distances between pairs of genomes, depending on the genome model and the allowed set of rearrangements. For instance, if one considers reversals in unsigned permutations, the problem is known to be NP-hard [4], while it is polynomial-time solvable in signed permutations [10]. We refer the reader to Fertin et al.'s book [7] for a survey of the algorithmics aspects of the subject.

As previously mentioned, almost all of these works have so far assumed that a genome is an ordered sequence of genes. However, it has recently been argued that this model could underestimate the "true" evolutionary distance, and that other genome features may require to be taken into account in the model in order to circumvent this problem [1,2].

Indeed, genomes carry more information than just their ordered sequences of genes, and in particular consecutive genes in a genome are separated by *intergenic regions*, which are DNA sequences between genes having different sizes (in terms of number of nucleotides).

This recently led some authors to model a genome as an ordered sequence of genes, together with an ordered list of its intergenic sizes, and to consider the problem of computing the DCJ distance, either in the case where insertions and deletions of nucleotides are forbidden [6], or allowed [3].

In this work, we also consider genomes as ordered sequences of genes together with their intergenic sizes, in the case where the gene sequence is an unsigned permutation and where the considered rearrangement operation is *super short reversal* (or SSR, i.e. a reversal of (gene) length at most two). In this context, our goal is to determine the minimum number of SSRs that transform one genome into another.

Sorting by super short reversals and/or super short transpositions (i.e. transpositions of (gene) length at most two each) has been studied in linear and circular genomes, signed and unsigned, and in all cases the problem has been shown to be in P class [8,9,11–13].

This paper is organized as follows. In Sect. 2 we provide the notations that we will use throughout the paper, and we introduce new notions that will prove useful for studying the problem. In Sect. 3, we derive lower and upper bounds on the sought distance, which in turn will help us design three different algorithms: one applies to the general case, while the remaining two apply to specific classes of genomes. Section 4 concludes the paper.

2   Definitions

We can represent a genome $\mathcal{G}$ with n genes as an n-tuple. When there is no duplicated genes, the n-tuple is a permutation $\pi = (\pi_1 \ \pi_2 \ ... \ \pi_{n-1} \ \pi_n)$ with $\pi_i \in \{1, 2, ..., (n-1), n\}$, for $1 \le i \le n$, and $\pi_i = \pi_j$ if, and only if, $i = j$. We denote by ι the *identity permutation*, the permutation in which all elements are in ascending order. The *extended permutation* is obtained from π by adding two new elements: $\pi_0 = 0$ and $\pi_{n+1} = (n+1)$.

A genome $\mathcal{G}$, represented by a permutation π with n elements, has $m = n+1$ intergenic regions $r^\pi = (r_1^\pi, ..., r_m^\pi)$, with $r_j^\pi \ge 0$ for $1 \le j \le m$, such that the intergenic region r_i^π is located before element π_i, for $1 \le i \le n$, and the intergenic region r_m^π is situated right after element π_n.

A *reversal* $\rho(i, j, x, y)$ applied over a permutation π, with $1 \le i \le j \le n$, $0 \le x \le r_i^\pi$, and $0 \le y \le r_{j+1}^\pi$, is an operation that (i) reverses the order of the elements in the subset of adjacent elements $\{\pi_i, ..., \pi_j\}$; (ii) reverses the order of intergenic regions in the subset of adjacent intergenic regions $\{r_{i+1}^\pi, ..., r_j^\pi\}$ when $j > i+2$; (iii) *cuts* two intergenic regions: after position x inside intergenic region r_i^π and after position y inside intergenic region r_{j+1}^π. This reversal results in the permutation π' such that $r_i^{\pi'} = x + y$ and $r_{j+1}^{\pi'} = (r_i^\pi - x) + (r_{j+1}^\pi - y)$.

A reversal $\rho(i, j, x, y)$ is also called a k-reversal, where $k = (j-i)+1$. A *super short reversal* is a 1-reversal or a 2-reversal, i.e., a reversal that affects only one or two elements of π.

Figure 1 shows a sequence of three super short reversals that transforms the permutation $\pi = (1 \ 3 \ 4 \ 2 \ 5)$ with $r^\pi = (3, 5, 2, 1, 2, 8)$ into $\iota = (1 \ 2 \ 3 \ 4 \ 5)$ with $r^\iota = (3, 2, 6, 4, 5, 1)$.

A pair of elements (π_i, π_j) from π is called an *inversion* if $\pi_i > \pi_j$ and $i < j$, with $\{i, j\} \in [1..n]$. We denote the number of inversions in a permutation π by $inv(\pi)$. For the example above, $inv(\pi) = 2$.

Given two permutations π and α of same size, representing genomes $\mathcal{G}_1$ and $\mathcal{G}_2$ respectively, we denote by $W_i(\pi, \alpha) = r_i^\pi - r_i^\alpha$ the *imbalance* between intergenic regions r_i^π and r_i^α, with $1 \le i \le m$.

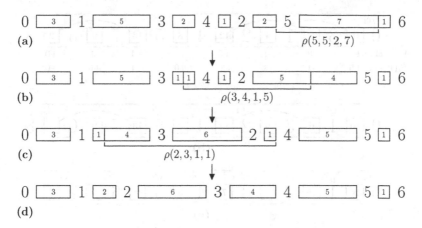

Fig. 1. A sequence of super short reversals that transforms $\pi = (1\ 3\ 4\ 2\ 5)$, with $r^\pi = (3, 5, 2, 1, 2, 8)$ into $\iota = (1\ 2\ 3\ 4\ 5)$, with $r^\iota = (3, 2, 6, 4, 5, 1)$. Intergenic regions are represented by rectangles, whose dimensions vary according to their sizes. The 1-reversal $\rho(5, 5, 2, 7)$ applied in **(a)** transforms π into $\pi' = \pi$, and it cuts π after position 2 at r_5^π and after position 7 at r_6^π, resulting in $r_5^{\pi'} = 9$, $r_6^{\pi'} = 1$, and $r^{\pi'} = (3, 5, 2, 1, 9, 1)$. The 2-reversal $\rho(3, 4, 1, 5)$ applied in **(b)** transforms π' into $\pi'' = (1\ 3\ 2\ 4\ 5)$, and it cuts π' after position 1 at $r_3^{\pi'}$ and after position 5 at $r_5^{\pi'}$, resulting in $r_3^{\pi''} = 6$, $r_5^{\pi''} = 5$, and $r^{\pi''} = (3, 5, 6, 1, 5, 1)$. Finally, the 2-reversal $\rho(2, 3, 1, 1)$ applied in **(c)** transforms π'' into ι, as shown in **(d)**.

Given two permutations π and α of same size and same total sum of the intergenic region lengths, let $S_j(\pi, \alpha) = \sum_{i=1}^{j} W_i(\pi, \alpha)$ be the cumulative sum of imbalances between intergenic regions of π and α from position 1 to j, with $1 \leq j \leq m$. Since π and α have same total sum of the intergenic region lengths, $S_m(\pi, \alpha) = 0$.

From now on, we will consider that (i) the target permutation α is such that $\alpha = \iota$; (ii) π and ι have the same number of elements; and (iii) the number of nucleotides inside intergenic regions of r^π equals the number of nucleotides inside intergenic regions of r^ι. By doing this, we can compute the *sorting distance* of π, denoted by $d(\pi)$, that consists in finding the minimum number of super short reversals that sorts π and transforms r^π into r^ι.

The *intergenic graph* of π with respect to the target permutation ι, denoted by $I(\pi, \iota) = (V, E)$, is such that V is composed by the set of intergenic regions r^π and the set of elements from the extended permutation π. Besides, the edge $e = (r_i^\pi, r_{i+2}^\pi) \in E$ if there is a $j \neq i$ such that (π_i, π_j) or (π_j, π_{i+1}) is an inversion, with $1 \leq i \leq n-1$ and $1 \leq j \leq n$.

A *component* c is a minimal set of consecutive elements from V in which: (i) the sum of imbalances of its intergenic regions with respect to r^ι is equal to zero; and (ii) any two intergenic regions that are connected to each other by an edge must belong to the same component.

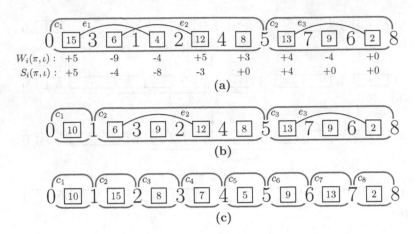

Fig. 2. Intergenic graphs $I(\pi, \iota)$ in **(a)**, $I(\pi', \iota)$ in **(b)**, and $I(\iota, \iota)$ in **(c)**, with $\pi = (3\ 1\ 2\ 4\ 5\ 7\ 6)$, $r^\pi = (15, 6, 4, 12, 8, 13, 9, 2)$, $\pi' = (1\ 3\ 2\ 4\ 5\ 7\ 6)$, $r^{\pi'} = (10, 6, 9, 12, 8, 13, 9, 2)$, $\iota = (1\ 2\ 3\ 4\ 5\ 6\ 7)$, and $r^\iota = (10, 15, 8, 7, 5, 9, 13, 2)$. Black squares represent intergenic regions, and the number inside it indicate their sizes. Rounded rectangles in blue represent components. Note that in **(a)** there are three edges in $I(\pi, \iota)$, and $C(I(\pi, \iota)) = 2$. We also have in **(a)** all values for $S_i(\pi, \iota)$ and $W_i(\pi, \iota)$, with $1 \leq i \leq 8$. The permutation π' is the result of applying $\rho(1, 2, 8, 2)$ to π. In **(b)** we can see that $I(\pi', \iota)$ has one more component than $I(\pi, \iota)$, and the edge e_1 was removed. In **(c)** we can see that when we reach the target permutation the number of components is equal to the number of intergenic regions in ι (i.e., $C(I(\iota, \iota)) = m = 8$).

A component always starts and finishes with elements from π. Besides, the first component starts with the element π_0, and the last component ends with the element π_{n+1}. Consecutive components share exactly one element from π, i.e., the last element π_i of a component is the first element of its adjacent component to the right. A component with one intergenic region is called *trivial*. The number of intergenic regions in a component c is denoted by $r(c)$. The number of components in a permutation π is denoted by $C(I(\pi, \iota))$. Figure 2 shows three examples of intergenic graphs.

3   Sorting Permutations by Super Short Reversals

In this section we analyze the version of the problem when only super short reversals (i.e., 1-reversals and 2-reversals) are allowed to sort a permutation on both order and intergenic regions. First, we show that any 1-reversal can increase the number of components by no more than one unit. After that, we state that if a component c of an intergenic graph $I(\pi, \iota)$ with $r(c) > 1$ has no edges (i.e., there is no inversions inside c), then it is always possible to split c into two components with a 1-reversal.

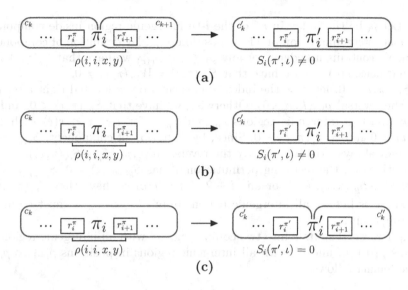

Fig. 3. Example of intergenic graphs for all possible values of $C(I(\pi', \iota))$ with respect to $C(I(\pi, \iota))$, where π' is the resulting permutation after applying a 1-reversal $\rho(i, i, x, y)$ to π. If the 1-reversal is applied over two components at the same time and $x + y \neq r_i^\pi$, then $C(I(\pi', \iota)) = C(I(\pi, \iota)) - 1$, as shown in **(a)**. If the 1-reversal is applied over one component, then either $C(I(\pi', \iota)) = C(I(\pi, \iota))$, if $x + y \neq r_i^\pi - S_i(\pi, \iota)$, or $C(I(\pi', \iota)) = C(I(\pi, \iota)) + 1$, if $x + y = r_i^\pi - S_i(\pi, \iota)$, as shown in **(b)** and **(c)** respectively.

Lemma 1. *Given a permutation π and a target permutation ι, let π' be the resulting permutation from π after applying a 1-reversal. It follows that $C(I(\pi, \iota)) - 1 \leq C(I(\pi', \iota)) \leq C(I(\pi, \iota)) + 1$.*

Proof. If a 1-reversal $\rho(i, i, x, y)$, applied over intergenic regions r_i^π and r_{r+1}^π, is applied over two different components in $I(\pi, \iota) = (V, E)$, then r_i^π is the last element of the first component, so $S_i(\pi, \iota) = 0$ and the graph $I(\pi', \iota) = (V', E')$, where π' is the resulting permutation, is such that $C(I(\pi', \iota)) = C(I(\pi, \iota)) - 1$ if $x + y \neq r_i^\pi$, as shown in Fig. 3(a). Let us consider now that this 1-reversal is applied over intergenic regions of a same component c.

First note that, since 1-reversals does not remove inversions from π, the intergenic graph $I(\pi', \iota)$ has $E' = E$. If $(r_i^{\pi'}, r_{i+2}^{\pi'}) \in E'$ (for $0 < i < n$), or $(r_{i-1}^{\pi'}, r_{i+1}^{\pi'}) \in E'$ (for $0 < i \leq n$), then $C(I(\pi', \iota)) = C(I(\pi, \iota))$. Otherwise, we have two cases to consider: $C(I(\pi', \iota)) = C(I(\pi, \iota))$, if $S_i(\pi', \iota) \neq 0$ (as shown in Fig. 3(b)); and $C(I(\pi', \iota)) = C(I(\pi, \iota)) + 1$ if $S_i(\pi', \iota) = 0$ (as shown in Fig. 3(c)). ☐

Lemma 2. *If a component c of an intergenic graph $I(\pi, \iota)$ with $r(c) \geq 2$ contains no edges, then there is always a pair of consecutive intergenic regions to which we can apply a 1-reversal that splits c into two components c' and c'' such that $r(c') + r(c'') = r(c)$.*

Proof. Let p_i be the index in r^π of the i-th intergenic region inside component c. The last intergenic region of c is at position $p_{r(c)}$. By definition of component, and since c contains no edges, for any $p_1 \le j < p_{r(c)}$ we have that $S_j(\pi, \iota) \ne 0$. Note that since $r(c) > 1$ we have that $S_{p_1}(\pi, \iota) = W_{p_1}(\pi, \iota) \ne 0$.

If $S_{p_1}(\pi, \iota) > 0$, let k be the index of element from π located right after $r_{p_1}^\pi$. Apply the reversal $\rho(k, k, r_{p_1}^\iota, 0)$. Otherwise, we have that $S_{p_1}(\pi, \iota) < 0$, and we need to find two intergenic regions $r_{p_i}^\pi$ and $r_{p_{i+1}}^\pi$ for $1 \le i < r(c)$ such that $S_{p_i}(\pi, \iota) < 0$ and $S_{p_{i+1}}(\pi, \iota) \ge 0$. Since, by definition of component, $S_{p_{r(c)}} = 0$, such a pair always exists. So, apply the reversal $\rho(p_i, p_i, r_{p_i}^\pi, -S_{p_i}(\pi, \iota))$.

In both cases, the resulting permutation π' has $S_{p_i}(\pi', \iota) = 0$, $S_{p_{i+1}}(\pi', \iota) = S_{p_{i+1}}(\pi, \iota) + S_{p_i}(\pi, \iota)$, and for any $i + 2 \le j \le r(c)$ we have that $S_{p_j}(\pi', \iota) = S_{p_j}(\pi, \iota)$ so, as before, all intergenic regions from $r_{p_{i+1}}^{\pi'}$ to $r_{p_{r(c)}}^{\pi'}$ must be in the same component.

This 1-reversal splits c into two components: c' with all intergenic regions in positions p_1 to p_i, and c'' with all intergenic regions in positions p_{i+1} to $p_{r(c)}$, and the lemma follows.   □

Now we state that any 2-reversal can increase the number of components by no more than two units.

Lemma 3. *Given a permutation π and a target permutation ι, let π' be the resulting permutation from π after applying a 2-reversal. We have that $C(I(\pi, \iota)) - 2 \le C(I(\pi', \iota)) \le C(I(\pi, \iota)) + 2$.*

Proof. If a 2-reversal is applied over intergenic regions of two different components then we are necessarily creating a new inversion, and the graph $I(\pi', \iota) = (V', E')$, where π' is the resulting permutation, has $C(I(\pi', \iota)) = C(I(\pi, \iota)) - 2$ (as shown in Fig. 4(a)) or $C(I(\pi', \iota)) = C(I(\pi, \iota)) - 1$ (as shown in Fig. 4(b)). Let us consider now that the operation is applied over intergenic regions of a same component c.

Suppose that we apply an operation that exchanges elements π_i and π_{i+1}, with $1 \le i < n - 1$. If the resulting permutation π' is such that $(r_i^{\pi'}, r_{i+2}^{\pi'}) \in E'$ then $C(I(\pi', \iota)) = C(I(\pi, \iota))$. Otherwise, we have three cases to consider: $C(I(\pi', \iota)) = C(I(\pi, \iota))$, if $S_i(\pi', \iota) \ne 0$ and $S_{i+1}(\pi', \iota) \ne 0$ (as shown in Fig. 4(c)); $C(I(\pi', \iota)) = C(I(\pi, \iota)) + 1$ if either $S_i(\pi', \iota) = 0$ or $S_{i+1}(\pi', \iota) = 0$ (as shown in Fig. 4(d)); and $C(I(\pi', \iota)) = C(I(\pi, \iota)) + 2$ otherwise (as shown in Fig. 4(e)).   □

Using Lemmas 1, 2, and 3 we show in the following two lemmas the minimum and maximum number of super short reversals needed to transform π into ι and r^π into r^ι.

Lemma 4. *Given a genome $\mathcal{G}_1$, let π be its corresponding permutation with $r^\pi = (r_1^\pi, ..., r_m^\pi)$ intergenic regions. We have that $d(\pi) \ge \max(\frac{m - C(I(\pi, \iota))}{2}, inv(\pi))$, where ι is the corresponding permutation of the target genome $\mathcal{G}_2$.*

Proof. In order to sort π we need to remove all inversions, and since a 2-reversal can remove only one inversion, we necessarily have that $d(\pi) \ge inv(\pi)$. Besides,

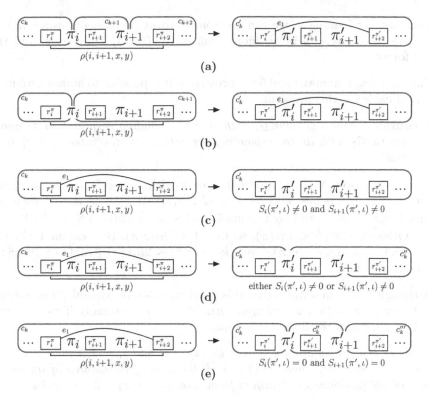

Fig. 4. Example of intergenic graphs for all possible values of $C(I(\pi', \iota))$ with respect to $C(I(\pi, \iota))$ where π' is the resulting permutation after applying a 2-reversal to π. When the 2-reversal is applied over two components at the same time then either $C(I(\pi', \iota)) = C(I(\pi, \iota)) - 2$, as shown in **(a)**, or $C(I(\pi', \iota)) = C(I(\pi, \iota)) - 1$, as shown in **(b)**. Otherwise, we have that either $C(I(\pi', \iota)) = C(I(\pi, \iota))$, if $S_i(\pi', \iota) \neq 0$ and $S_{i+1}(\pi', \iota) \neq 0$ as shown in **(c)**, or $C(I(\pi', \iota)) = C(I(\pi, \iota)) + 1$, if $e_1 \notin I(\pi', \iota)$ and either $S_i(\pi', \iota) \neq 0$ or $S_{i+1}(\pi', \iota) \neq 0$ as shown in **(d)**, or $C(I(\pi', \iota)) = C(I(\pi, \iota)) + 2$, if $e_1 \notin I(\pi', \iota)$, $S_i(\pi', \iota) = 0$ and $S_{i+1}(\pi', \iota) = 0$ as shown in **(e)**.

by Lemmas 1 and 3, we can increase the number of components by at most two with a super short reversal, so to reach m trivial components we need at least $\frac{m - C(I(\pi, \iota))}{2}$ super short reversals. Thus, $d(\pi) \geq \max(\frac{m - C(I(\pi, \iota))}{2}, inv(\pi))$. $\square$

Lemma 5. *Given a genome $\mathcal{G}_1$, let π be its corresponding permutation with $r^\pi = (r_1^\pi, ..., r_m^\pi)$ intergenic regions. We have that $d(\pi) \leq inv(\pi) + m - C(I(\pi, \iota))$, where ι is the corresponding permutation of the target genome $\mathcal{G}_2$.*

Proof. Suppose that first we remove all inversions of π with $inv(\pi)$ 2-reversals of type $\rho(i, i+1, r_i^\pi, 0)$ i.e., without exchanging its intergenic regions. Let π' be the resulting permutation, with $r^{\pi'} = r^\pi$. The number of components in π cannot be smaller than $C(I(\pi, \iota))$ since each 2-reversal removing an inversion is applied inside a same component. Let us suppose then that π' has $k' \geq C(I(\pi, \iota))$

components. By Lemma 2, we can go from k' to m components using $m - k'$ 1-reversals, which results in no more than $m - C(I(\pi, \iota))$ 1-reversals, and the lemma follows.   □

Finally, using Lemmas 4 and 5, we prove that it is possible to obtain a solution 3-approximable for this problem.

Theorem 6. *Given a genome $\mathcal{G}_1$ with its corresponding permutation π, and a target genome $\mathcal{G}_2$ with its corresponding permutation ι, the value of $d(\pi)$ is 3-approximable.*

Proof. Let us represent $\mathcal{G}_1$ by a permutation π with $r^\pi = (r_1^\pi, ..., r_m^\pi)$ intergenic regions, $inv(\pi)$ inversions, and let $k = C(I(\pi, \iota))$. If $\frac{m-k}{2} > inv(\pi)$ then, by Lemma 4, $d(\pi) \geq \frac{m-k}{2}$, and, by Lemma 5, $d(\pi) \leq m-k+inv(\pi) \leq m-k+\frac{m-k}{2} \leq 3\frac{m-k}{2}$. Otherwise, $\frac{m-k}{2} < inv(\pi)$, so $m - k < 2inv(\pi)$. By Lemma 4, $d(\pi) \geq inv(\pi)$, and, by Lemma 5, $d(\pi) \leq m - k + inv(\pi) \leq 2inv(\pi) + inv(\pi) \leq 3inv(\pi)$, and the lemma follows.   □

Although Theorem 6 states that this problem is 3-approximable, it is possible to sort any permutation π and transform r^π into r^ι optimally if $\pi_1 = n$ and $\pi_n = 1$, as shown in the following lemma.

Lemma 7. *If a permutation π is such that $\pi_1 = n$ and $\pi_n = 1$, with $n > 1$, then $d(\pi) = inv(\pi) + \varphi(\pi)$, where $\varphi(\pi) = 1$, if the sum of imbalances of intergenic regions in odd positions of r^π differs from zero, and $\varphi(\pi) = 0$, otherwise.*

Proof. By Lemma 4, we have that $d(\pi) \geq inv(\pi)$. Besides, since only 2-reversals remove inversions, and since 2-reversals exchange nucleotides between intergenic regions of same parity only, then $d(\pi) \geq inv(\pi) + \varphi(\pi)$, with $\varphi(\pi) = 1$, if the cumulative sum of imbalances of intergenic regions in odd positions, denoted by $S_{odd}(\pi, \iota)$, differs from zero (in this case we will need at least one 1-reversal to exchange nucleotides between an odd and an even intergenic region), and $\varphi(\pi) = 0$ otherwise. Consider the following procedure, divided into four steps:

(i) Remove any inversion between elements in positions 2 to $(n - 1)$ with 2-reversals of type $\rho(i, i + 1, r_i^\pi, 0)$, and let $\pi' = (n\ 2\ ...\ (n-1)\ 1)$ be the resulting permutation. Note that $r^{\pi'} = r^\pi$, and π' has $(2n - 3)$ inversions which means that $inv(\pi) - 2n + 3$ 2-reversals were applied.

(ii) Take the element $\pi'_1 = n$ to position $n - 1$ by a sequence of $(n-2)$ 2-reversals of type $\rho(i, i+1, 0, 0)$, for $1 \leq i \leq n-2$, and let $\pi'' = (2\ 3\ ...\ n\ 1)$ be the resulting permutation. After this sequence is applied, all intergenic nucleotides are in the last three intergenic regions $r_{n-1}^{\pi''}, r_n^{\pi''}$ and $r_{n+1}^{\pi''}$ only, and $inv(\pi'') = n - 1$.

(iii) Let $a = S_{odd}(\pi'', \iota)$, if n is odd, and $a = -S_{odd}(\pi, \iota)$ otherwise, and let $b = W_{n+1}(\pi'', \iota)$. If $b \geq 0$ (resp. $b < 0$) apply the 2-reversal $\rho(n-1, n, r_{n-1}^{\pi''}, b)$ balancing r_{n+1} (resp. if $a \neq 0$, apply the 1-reversal $\rho(n-1, n-1, x, y)$ with $x = r_{n-1}^{\pi''}$ and $y = a$ if $a > 0$; $x = r_{n-1}^{\pi''}+a$ and $y = 0$ otherwise), and, if $a \neq 0$, apply $\rho(n-1, n-1, x, y)$, with $x = r_{n-1}^{\pi''}+b$ and $y = a$ if $a > 0$; $x = r_{n-1}^{\pi''}+b+a$

and $y = 0$ otherwise (resp. apply by the 2-reversal $\rho(n-1, n, x + y + b, 0)$ balancing r_{n+1}). We applied $1+\varphi(\pi)$ operations here. Let $\pi''' = (2 \ ... \ 1 \ n)$ be the resulting permutation, with $(n-2)$ inversions and two components: one with all intergenic regions $r_i^{\pi'''}$, for $1 \leq i \leq n$, and one with the intergenic region $r_{n+1}^{\pi'''}$ only.

(iv) Move element 1 from position $(n-1)$ to position 1 by a sequence of reversals $\rho(i, i + 1, 0, k - r_{i+2}^t)$ such that k is the length of the intergenic region that the current 2-reversal is cutting in the right. We will apply $(n - 2)$ 2-reversals, removing the same amount of inversions. This step goes from 2 to $2 + (n - 1) = m$ components since each 2-reversal here creates a new component, except for the last one that creates two new components.

Summing up, we apply $inv(\pi) - 2n + 3$ reversals in (i), $n - 2$ reversals in (ii), $1 + \varphi(\pi)$ reversals in (iii), and $n - 2$ reversals in (iv), which gives us exactly the minimum amount of $(inv(\pi) + \varphi(\pi))$ operations.   □

We can use Lemma 7 to obtain a 2-approximation algorithm for permutations π with $n \geq 9$ elements and $inv(\pi) \geq 4n$, as explained in the next lemma.

Lemma 8. *If a permutation π with $n \geq 9$ elements has $inv(\pi) \geq 4n$ then the value of $d(\pi)$ is 2-approximable.*

Proof. Suppose that we have a permutation π with $n \geq 9$ such that $inv(\pi) \geq 4n$. By Lemma 4, we have that $d(\pi) \geq inv(\pi)$. Consider the following procedure, divided into three steps:

(i) Apply a sequence of k super short reversals that moves the element n on π to position 1, without exchanging any intergenic region (i.e., any super short reversal $\rho(i, i+1, x, y)$ applied here has $x = r_i^\pi$ and $y = 0$, keeping r^π intact). Let π' be the resulting permutation. Since π has n elements, we have that $k < n$ and $inv(\pi') < inv(\pi) + n$, regardless of the position of element n in π.

(ii) Apply a sequence of k' super short reversals in a similar way as above that moves element 1 from π' to position n. Let π'' be the resulting permutation. Since π' has n elements, and since element 1 cannot be at position 1 in π' ($\pi_1' = n$), it follows that $k' < n - 1$ and $inv(\pi'') < inv(\pi') + n - 1 < inv(\pi) + 2n - 1$, regardless of the position of element 1 in π'.

(iii) Use the algorithm presented in Lemma 7 to sort π''.

Note that the first two steps apply $(k + k') < (2n - 1)$ super short reversals, and Step (iii) applies up to $inv(\pi) + 2n$ super short reversals, so the procedure above applies z super short reversals such that $z \leq 2n-1+inv(\pi)+2n = inv(\pi)+4n-1$. Since $inv(\pi) \geq 4n$, we have that $z \leq 2inv(\pi)$, and the lemma follows.   □

4   Conclusion

In this paper, we analyzed the minimum number of super short reversals needed to sort a permutation π and transform its intergenic regions r^π according to the set of intergenic regions r^ι of the target genome represented by ι. We defined some bounds that allowed us to state three different algorithms: a more general that guarantees an approximation factor of 3; an exact algorithm for any permutation π with $n > 1$ elements such that $\pi_1 = n$ and $\pi_n = 1$; and a more specific one that sorts any permutation π with $n \geq 9$ elements such that $inv(\pi) \geq 4n$ with an approximation factor of 2. We intend to investigate the problem using super short transpositions instead of super short reversals, as well as using these operations together on signed permutations. We will also study the complexity of all these variants of the problem.

Acknowledgments. This work was supported by the National Council for Scientific and Technological Development - CNPq (grants 400487/2016-0, 425340/2016-3, and 140466/2018-5), the São Paulo Research Foundation - FAPESP (grants 2013/08293-7, 2015/ 11937-9, 2017/12646-3, 2017/16246-0, and 2017/16871-1), the Brazilian Federal Agency for the Support and Evaluation of Graduate Education - CAPES, and the CAPES/COFECUB program (grant 831/15).

References

1. Biller, P., Guéguen, L., Knibbe, C., Tannier, E.: Breaking good: accounting for fragility of genomic regions in rearrangement distance estimation. Genome Biol. Evol. **8**(5), 1427–1439 (2016)
2. Biller, P., Knibbe, C., Beslon, G., Tannier, E.: Comparative genomics on artificial life. In: Beckmann, A., Bienvenu, L., Jonoska, N. (eds.) CiE 2016. LNCS, vol. 9709, pp. 35–44. Springer, Cham (2016). https://doi.org/10.1007/978-3-319-40189-8_4
3. Bulteau, L., Fertin, G., Tannier, E.: Genome rearrangements with indels in intergenes restrict the scenario space. BMC Bioinform. **17**(S14), 225–231 (2016)
4. Caprara, A.: Sorting permutations by reversals and eulerian cycle decompositions. SIAM J. Discret. Math. **12**(1), 91–110 (1999)
5. Chen, T., Skiena, S.S.: Sorting with fixed-length reversals. Discret. Appl. Math. **71**(1–3), 269–295 (1996)
6. Fertin, G., Jean, G., Tannier, E.: Algorithms for computing the double cut and join distance on both gene order and intergenic sizes. Algorithms Mol. Biol. **12**(16), 1–11 (2017)
7. Fertin, G., Labarre, A., Rusu, I., Tannier, E., Vialette, S.: Combinatorics of Genome Rearrangements. Computational Molecular Biology. The MIT Press, London (2009)
8. Galvão, G.R., Baudet, C., Dias, Z.: Sorting circular permutations by super short reversals. IEEE/ACM Trans. Comput. Biol. Bioinform. **14**(3), 620–633 (2017)
9. Galvão, G.R., Lee, O., Dias, Z.: Sorting signed permutations by short operations. Algorithms Mol. Biol. **10**(12), 1–17 (2015)
10. Hannenhalli, S., Pevzner, P.A.: Transforming men into mice (polynomial algorithm for genomic distance problem). In: Proceedings of the 36th Annual Symposium on Foundations of Computer Science (FOCS 1995), pp. 581–592. IEEE Computer Society Press, Washington, DC (1995)

11. Jerrum, M.R.: The complexity of finding minimum-length generator sequences. Theor. Comput. Sci. **36**(2–3), 265–289 (1985)
12. Knuth, D.E.: The art of Computer Programming: Fundamental Algorithms. Addison-Wesley, Reading (1973)
13. Oliveira, A.R., Fertin, G., Dias, U., Dias, Z.: Sorting signed circular permutations by super short operations. Algorithms Mol. Biol. **13**(13), 1–16 (2018)
14. Yancopoulos, S., Attie, O., Friedberg, R.: Efficient sorting of genomic permutations by translocation, inversion and block interchange. Bioinformatics **21**(16), 3340–3346 (2005)

Identifying Maximal Perfect Haplotype Blocks

Luís Cunha[1,2], Yoan Diekmann[3], Luis Kowada[1], and Jens Stoye[1,4(✉)]

[1] Universidade Federal Fluminense, Niterói, Brazil
[2] Universidade Federal do Rio de Janeiro, Rio de Janeiro, Brazil
[3] Department of Genetics, Evolution and Environment, University College London, London WC1E 6BT, UK
[4] Faculty of Technology and Center for Biotechnology, Bielefeld University, Bielefeld, Germany
jens.stoye@uni-bielefeld.de

Abstract. The concept of maximal perfect haplotype blocks is introduced as a simple pattern allowing to identify genomic regions that show signatures of natural selection. The model is formally defined and a simple algorithm is presented to find all perfect haplotype blocks in a set of phased chromosome sequences. Application to three whole chromosomes from the 1000 genomes project phase 3 data set shows the potential of the concept as an effective approach for quick detection of selection in large sets of thousands of genomes.

Keywords: Population genomics · Selection coefficient
Haplotype block

1 Introduction

Full genome sequences are amassing at a staggering yet further accelerating pace. For humans, multiple projects now aim to deliver numbers five orders of magnitudes higher than the initial human genome project 20 years ago[1]. This development is fuelled by drastic reductions in sequencing costs, by far exceeding Moore's law [6]. As a result, the bottleneck in genomics is shifting from data production to analysis, calling for more efficient algorithms scaling up to ever-larger problems.

A simple yet highly interesting pattern in population genomic datasets are fully conserved haplotype blocks (called *maximal perfect haplotype blocks* in the following). When large and frequent enough in the population, they may be indicative for example of a selective sweep. Their simple structure simplifies analytical treatment in a population genetic framework. To our surprise, we could not locate any software tool that can find them efficiently.

---

[1] *E.g.*, https://www.genomicsengland.co.uk/the-100000-genomes-project-by-numbers.

R. Alves (Ed.): BSB 2018, LNBI 11228, pp. 26–37, 2018.
https://doi.org/10.1007/978-3-030-01722-4_3

In this paper, we present a simple and efficient algorithm for finding all maximal perfect haplotype blocks in a set of binary sequences. While the algorithm is not optimal in terms of worst-case analysis, and therefore this paper ends with an interesting open question, it has the convenient property allowing its *lazy* implementation, making it applicable to large real data sets in practice.

The paper is organized as follows: The problem of finding all maximal perfect haplotype blocks in a set of binary sequences is formally defined in Sect. 2. The trie data structure we use and our algorithm are presented in Sect. 3. In Sect. 4 we describe our lazy implementation of the algorithm and show its applicability to real data. Section 5 concludes with an open problem and two ideas for alternative algorithmic approaches.

2   Basic Definitions

The input data to our problem are k binary sequences, all of the same length n, each one representing a chromosome. Each column refers to a biallelic[2] single nucleotide polymorphism (SNP), with entries 0 and 1 corresponding to different but otherwise arbitrary alleles, although polarised data (where 0 and 1 refer to ancestral and derived allele, respectively) usually helps the interpretation of results. Of special interest are large blocks of conservation, that are stretches of identical alleles present at the same loci in many of the input sequences. Formally, we define such blocks as follows.

Definition 1. *Given an ordered set $X = (x_1, \ldots, x_k)$ and an index set $I = \{i_1, \ldots, i_\ell\}$, $0 \leq \ell \leq k$, $1 \leq i_j \leq k$, the I-induced subset of X is the set $X|_I = \{x_{i_1}, \ldots, x_{i_\ell}\}$.*

Definition 2. *Given k binary sequences $S = (s_1, \ldots, s_k)$, each of length n, a maximal haplotype block is a triple (K, i, j) with $K \subseteq \{1, \ldots, k\}$, $|K| \geq 2$ and $1 \leq i \leq j \leq n$ such that*

1. *$s[i..j] = t[i..j]$ for all $s, t \in S|_K$,*
2. *$i = 1$ or $s[i-1] \neq t[i-1]$ for some $s, t \in S|_K$ (left-maximality),*
3. *$j = n$ or $s[j+1] \neq t[j+1]$ for some $s, t \in S|_K$ (right-maximality), and*
4. *there exists no $K' \subseteq \{1, \ldots, k\}$, $K' \supsetneq K$, such that $s[i..j] = t[i..j]$ for all $s, t \in S|_{K'}$.*

The formal problem we address in this paper can then be phrased as follows.

Problem 1. Given k binary sequences $S = (s_1, \ldots, s_k)$, each of length n, find all maximal haplotype blocks in S.

---

[2] For convenience, we exclude multiallelic sites which may contain alleles coded as 2 or 3, or merge the minor alleles if they are rare and represent them as 1. These make up only a small fraction of the total SNPs in real data, and we therefore do not expect any overall effect.

The following proposition gives a simple upper bound for the output of this problem.

Proposition 1. *Given k binary sequences $S = (s_1, \ldots, s_k)$, each of length n, there can be only $O(kn)$ maximal haplotype blocks in S.*

Proof. We argue that at any position i, $1 \le i \le n$, there can start at most $k - 1$ maximal haplotype blocks. This follows from the maximality condition and the fact that a maximal haplotype block contains at least two sequences whose longest common prefix it is.   □

As the following example shows, for sufficiently large n the bound given in Proposition 1 is tight.

Example 1. Consider the family of sequences $S_{k,n} = (s_1, s_2, \ldots, s_k)$, each of length n, defined as follows: $s_1 = 0^n$, $s_2 = 1^n$, followed by *chunks* of sequences $c_1, c_2, \ldots$ such that chunk c_i contains $2i$ sequences, that are all sequences of length n which are repetitions of $0^i 1^i$ and its rotations. The last chunk may be truncated, so that the total number of sequences is k. For example, for $k = 14$ and $n = 24$ we have:

$$s_1 = 000000000000000000000000$$
$$s_2 = 111111111111111111111111$$

$$\left. \begin{array}{l} s_3 = 010101010101010101010101 \\ s_4 = 101010101010101010101010 \end{array} \right\} c_1$$

$$\left. \begin{array}{l} s_5 = 001100110011001100110011 \\ s_6 = 011001100110011001100110 \\ s_7 = 110011001100110011001100 \\ s_8 = 100110011001100110011001 \end{array} \right\} c_2$$

$$\left. \begin{array}{l} s_9 = 000111000111000111000111 \\ s_{10} = 001110001110001110001110 \\ s_{11} = 011100011100011100011100 \\ s_{12} = 111000111000111000111000 \\ s_{13} = 110001110001110001110001 \\ s_{14} = 100011100011100011100011 \end{array} \right\} c_3$$

Analysis shows that in chunk c_i, $i \ge 1$, every substring of length i forms a new maximal haplotype block together with some substring in one of the earlier sequences. Therefore, for $n > k$ the total number of maximal haplotype blocks in $S_{k,n}$ grows as $\Theta(kn)$.

3   Algorithm

Our algorithm to find all maximal haplotype blocks in a set of sequences S uses the (binary) trie of the sequences in S. For completeness, we recall its definition:

Definition 3. *The* trie *of a set of sequences S over an alphabet Σ is the rooted tree whose edges are labeled with characters from Σ and the labels of all edges starting at the same node are distinct, such that the concatenated edge labels from the root to the leaves spell exactly the sequences in S.*

A *branching vertex* in a rooted tree is a vertex with out-degree larger than one.

Example 2. Figure 1 shows the trie $T_1(S)$ of $k = 4$ binary strings of length $n = 6$. It has three branching vertices.

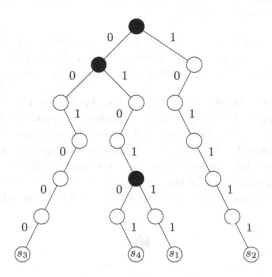

Fig. 1. Trie $T_1(S)$ of $k = 4$ binary strings $S = (s_1, s_2, s_3, s_4)$ with $s_1 = 010111$, $s_2 = 101111$, $s_3 = 001000$ and $s_4 = 010101$. Branching vertices are indicated by filled nodes.

It is well known that the trie of a set of sequences S over a constant-size alphabet can be constructed in linear time and uses linear space with respect to the total length of all sequences in S.

Our algorithm to find all maximal haplotype blocks of k binary sequences S, each of length n, iteratively constructs the trie of the suffixes of the sequences in S starting at a certain index i, $i = 1, 2, \ldots, n$. We denote the ith trie in this series of tries by $T_i(S)$.

Observation 1. *All branching vertices of $T_1(S)$ correspond to maximal haplotype blocks starting at index 1 of sequences in S.*

This follows from the fact that a branching vertex in $T_1(S)$ corresponds to a common prefix of at least two sequences in S that are followed by two different characters, thus they are right-maximal. Left-maximality is automatically given since $i = 1$.

Example 2. (cont'd). The tree $T_1(S)$ in Fig. 1 has three branching vertices, corresponding to the maximal haplotype blocks starting at index 1: the string 0101 occurring as a prefix in sequences s_1 and s_4; the string 0 occurring as a prefix in sequences s_1, s_3 and s_4; and the empty string (at the root of the tree) occurring as a prefix of all four strings.

In order to find maximal haplotype blocks that start at later positions $i > 1$, essentially the same idea can be used, just based on the tree $T_i(S)$. The only difference is that, in addition, one needs explicitly to test for left-maximality. As the following observation shows, this is possible by looking at the two subtrees of the root of the previous trie, $T_{i-1}(S)$.

Observation 2. *A haplotype block starting at position $i > 1$ is left-maximal if and only if it contains sequences that are in the 0-subtree of the root of $T_{i-1}(S)$ and sequences that are in the 1-subtree of the root of $T_{i-1}(S)$.*

Example 3. (cont'd). As shown in Fig. 2, trie $T_2(S)$ has three branching vertices, corresponding to the right-maximal haplotype blocks starting at index 2: the string 101 occurring at positions 2..4 in sequences s_1 and s_4; the string 01 occurring at positions 2..3 in sequences s_2 and s_3; and, again, the empty string. The string 101 is not left-maximal (and therefore not maximal), visible from the fact that s_1 and s_4 were both in the same (0-) subtree of the root in $S_1(T)$. The other two right-maximal haplotype blocks are also left-maximal.

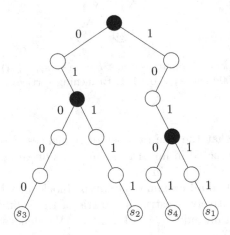

Fig. 2. Trie $T_2(S)$ for the strings from Fig. 1.

The algorithm to find all maximal haplotype blocks in a set of k sequences S, each of length n, follows immediately. It first constructs the trie $T_1(S)$ and locates all prefix haplotype blocks by a simple depth-first traversal. Then, iteratively for $i = 2, 3, \ldots, n$, $T_i(S)$ is constructed by merging the 1-subtree into the 0-subtree of the root of $T_{i-1}(S)$ during a parallel traversal of the two sister-subtrees. The former 0-subtree will then be $T_i(S)$. While doing so, branching vertices with leaves that came from both of these subtrees are reported as maximal blocks starting from index i. Pseudocode is given in Algorithm 1.

Algorithm 1. (Haploblocks)

Input: k binary sequences $S = (s_1, \ldots, s_k)$, each of length n
Output: all maximal haplotype blocks of S
 1: construct $T \leftarrow T_1(S)$
 2: **for each** branching vertex v of T **do**
 3:     report maximal block at positions $1..d - 1$, where d is the depth of v in T
 4: **end for**
 5: **for** $i = 2, \ldots, n$ **do**
 6:     merge-and-report(T.left, T.right; $i, 0$)
 7:     $T \leftarrow T$.left
 8: **end for**

Function merge-and-report$(l, r; i, d)$
 9: **if** l is empty **then**
10:     $l \leftarrow r$ ▷ simplification of presentation: implemented through *call by reference*
11:     **return**
12: **end if**
13: **if** r is empty **then**
14:     **return**
15: **end if**
16: leftmaximal $\leftarrow$ **not** (l.left and r.left are empty **or** l.right and r.right are empty)
17: **if** l.left is empty **then**
18:     l.left $\leftarrow r$.left
19: **else**
20:     merge-and-report(l.left, r.left; $i, d + 1$)
21: **end if**
22: **if** l.right is empty **then**
23:     l.right $\leftarrow r$.right
24: **else**
25:     merge-and-report(l.right, r.right; $i, d + 1$)
26: **end if**
27: rightmaximal $\leftarrow$ **not** (l.left is empty **or** l.right is empty)
28: **if** leftmaximal **and** rightmaximal **then**
29:     report maximal block at positions $i..i + d$
30: **end if**

Analysis. The overall running time of Algorithm 1 is $O(kn^2)$. This can be seen easily as follows. The initial construction of $T_1(S)$ takes linear time in the input size, thus $O(kn)$ time. Similar for the identification of maximal haplotype blocks starting at index $i = 1$. Each of the following $n - 1$ iterations performs in the worst case a traversal of the complete tree that has size $O(kn)$, thus taking $O(kn^2)$ time in total.

Note that, as presented in the pseudocode of Algorithm 1, the algorithm only reports the start and end positions i and j, respectively, of a maximal haplotype block (K, i, j), but not the set of sequences K where the block occurs. This can easily be added if, whenever in lines 3 and 29 some output is generated, the current subtree is traversed and the indices of all $|K|$ sequences found at the leaves are collected and reported. Such a traversal, however, costs $O(n \cdot |K|)$ time in the worst case, resulting in an overall running time of $O(kn^2 + n \cdot |\text{output}|)$. An alternative could be to store at each branching vertex of the trie as *witness* the index of a single sequence in the subtree below. This would allow to report, in addition to start and end positions i and j, respectively, also the sequence of a maximal haplotype block (K, i, j). If desired, the set K can then be generated easily using standard pattern matching techniques on the corresponding intervals of the k input sequences in $O(k \cdot (j - i))$ time.

4 Results

4.1 Data

To evaluate our algorithm, we downloaded chromosomes 2, 6 and 22 of the 1000 genomes phase 3 data set, which provides phased whole-genome sequences of 2504 individuals from multiple populations world-wide [1]. We extracted biallelic SNPs and represented the data as a binary matrix with help of the *cyvcf2* Python library [10].

4.2 Our Implementation of Algorithm 1

We implemented Algorithm 1 in C. Thereby we encountered two practical problems.

First, the recursive structure of Algorithm 1, when applied to haplotype sequences that are several hundred thousand characters long, produces a program stack overflow. Therefore we re-implemented the tree construction and traversal in a non-recursive fashion, using standard techniques as described, e.g., on the "Non-recursive depth first search algorithm" page of Stack Overflow[3].

Second, the constructed trie data structure requires prohibitive space. For example, already for the relatively small chromosome 22, $T_1(S)$ has 5,285,713,633 vertices and thus requires (in our implementation with 32 bytes per vertex) more than 157 gigabytes of main memory. However, most of the vertices are in non-branching paths to the leaves, corresponding to unique suffixes of sequences in S. Since such paths can never contain a branching vertex, they are not relevant. They become of interest only later in the procedure when the path is merged with other paths sharing a common prefix. Therefore we implemented a *lazy* version of our data structure, that stops the construction of a path whenever it contains only a single sequence. During the merge-and-report procedure, then, whenever an unevaluated part of the tree is encountered, the path has to be

---

[3] https://stackoverflow.com.

extended until it branches and paths represent single sequences again. This has the effect that at any time only the top part of the trie is explicitly stored in memory. For chromosome 22, the maximum number of nodes that are explicitly stored at once drops to 5,677,984, reducing the memory footprint by about a factor of 1,000. In fact, this number is not much larger for any other of the human chromosomes that we tested, since it depends on the size of the maximal perfect haplotype blocks present in the data, and not on the chromosome length.

Table 1 contains memory usage and running times for all three human chromosomes that we studied. All computations were performed on a Dell RX815 machine with 64 2.3 GHz AMD Opteron processors and 512 GB of shared memory.

Table 1. Resources used by our implementation of Algorithm 1 when applied to the three data sets described in Sect. 4.1.

Data set	Length	Memory	Time
chr. 2	6,786,300	33.67 GB	2 h 37 min
chr. 6	4,800,101	23.91 GB	1 h 51 min
chr. 22	1,055,454	5.45 GB	25 min

4.3 Interpretation of Results

In order to demonstrate the usefulness of the concept of haplotype blocks and our algorithm and implementation to enumerate them, we show how our results can form the efficient algorithmic core of a genome-wide selection scan.

Given a maximal perfect haplotype block (K, i, j) found in a set of k chromosomes, we estimate the selection coefficient s and the time t since the onset of selection following the approach presented by Chen et al. [3]. Therefore, we first convert the physical positions corresponding to indices i and j of the block from base pairs to a genetic distance d quantifying genetic linkage in centimorgan[4], which is the chromosomal distance for which the expected number of crossovers in a single generation is 0.01. Distance value d in turn is converted to the recombination fraction r – defined as the ratio of the number of recombined gametes between two chromosomal positions to the total number of gametes produced – using Haldane's map function

$$r = \frac{1 - \exp(-\frac{2d}{100})}{2}. \tag{1}$$

---

[4] A genetic map required to do so is available for example as part of Browning et al. [2] at http://bochet.gcc.biostat.washington.edu/beagle/genetic_maps.

With r, K, k we can define a likelihood function $\mathcal{L}(s \mid r, K, k)$ allowing to compute maximum likelihood estimates of the selection coefficient and time since the onset of selection, $\hat{s}$ and $\hat{t}$, respectively. The full derivation from population genetic theory is outside the scope of this paper, and the subsequent paragraphs merely intend to provide some basic intuition. For more details, we refer the interested reader to Chen et al. [3] and the appendix of reference [4].

First, assume a deterministic model for the frequency change of an allele with selective advantage s, which yields a sigmoidal function over continuous time t that ignores the stochasticity in frequency trajectories for small s,

$$y_t = \frac{y_0}{y_0 + (1 - y_0)e^{-st}}, \tag{2}$$

where y_0 is the initial allele frequency at the onset of selection[5] and $y_t = \frac{|K|}{k}$ is the observed allele frequency assumed to be representative of the population frequency. Equation 2 links up the selection coefficient s and the age t of the allele, for example requiring larger selective advantage to reach a given frequency if the allele is young.

Next, we exploit the fact that the recombination rate is independent of selection, and if assumed to be constant through time can therefore be seen to behave as a "recombination clock". Given a haplotype of length such that its recombination fraction is r, moreover, with an allele at one end that at time t segregates at frequency $y(t)$ in the population, the expected number of recombination events C altering that allele in the time interval $[0, t]$ can be obtained by

$$C = r \int_{u=0}^{t} (1 - y(u))\, du = r\left(t - \frac{1}{s}\ln(1 - y_0 + e^{st}y_0)\right), \tag{3}$$

where the second equality follows from substituting in Eq. 2. Assuming that the number of recombination events follows a Poisson distribution, the probability of no event, i.e. of full conservation of a haplotype after time t, becomes

$$e^{-C} = e^{-rt}(1 - y_0(1 - e^{st}))^{\frac{r}{s}}. \tag{4}$$

Finally, one can define the likelihood of observing a haplotype block for a given s and t as $|K|$ times the probability of a conserved haplotype (Eq. 4) times the probability of recombination events at the borders (Eq. 3). As usual, the logarithm simplifies the equation, yielding

$$\ln \mathcal{L}(s|r, K, k) \propto$$

$$- rt + \frac{r}{s}\ln(1 - y_0(1 - e^{st})) + \ln\left(t - \frac{1}{s}\ln(1 - y_0(1 - e^{st}))\right), \tag{5}$$

with t being directly derived from Eq. 2:

$$t = \frac{1}{s}\ln\left(\frac{y_t(1 - y_0)}{y_0(1 - y_t)}\right). \tag{6}$$

---

[5] In the following, y_0 is arbitrarily fixed at 0.00005, corresponding to $\frac{1}{2N_e}$ with an effective population size $N_e = 10,000$.

Note that we omit the factor $|K|$ and summands $\ln(r)$ for the recombination fractions at the borders of the haplotype (see Eq. 3) that we assume are small and approximately equal, as they are inconsequential for optimization. Also, the massive speed gain of our approach trades off with a systematic but conservative underestimation of $\hat{s}$ when compared to the original equation in reference [3] as we do not consider the full varying extent of the individual haplotypes.

Equation 5 can be evaluated for a range of values to find the (approximate) maximum likelihood estimate $\hat{s}$ at a given precision, e.g. $s \in \{0.001, 0.002, \dots\}$ to estimate $\hat{s}$ with error below 0.001. Once $\hat{s}$ has been found, the corresponding time $\hat{t}$ is obtained by substituting $\hat{s}$ into Eq. 6.

As the Haploblocks algorithm is able to rapidly scan entire chromosomes, and estimating $\hat{s}$ and $\hat{t}$ requires to evaluate only simple analytical expressions, one can efficiently generate a genome-wide selection track. Figure 3 illustrates the results for the locus known to contain one of the strongest signals of selection detected so far, the lactase persistence allele in modern Europeans -13.910:C>T (rs4988235). The selection coefficient we compute is consistent with the range of current estimates (see Ségurel and Bon [11] and references therein).

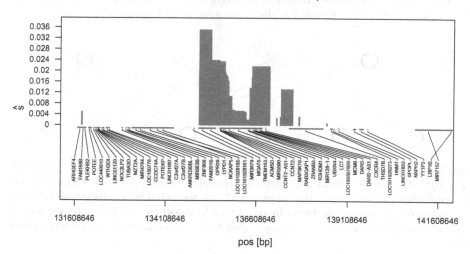

Fig. 3. Maximum likelihood estimates of selection coefficients for the locus containing the lactase gene in European individuals from the 1000 genomes data set. Each block of weight above 500,000 was converted to a selection coefficient applying Eq. 5 on a set of values $\{0.0002, 0.0004, \dots\}$ and choosing the (approximate) maximum likelihood estimate $\hat{s}$. Red lines indicate genes annotated in the RefSeq database [9]. (color figure online)

5 Conclusion

We presented an $O(kn^2)$ time algorithm for finding all maximal perfect haplotype blocks in a set of k binary sequences, each of length n, that scales well in practice. Even large human chromosomes can be processed in a few hours using moderate amounts of memory.

This allowed us to design an analytical approach with enumeration of maximal perfect haplotype blocks at its core that not only detects selection genome-wide efficiently, but does so by directly estimating a meaningful and interpretable parameter, the selection coefficient s.

As a proof of principle, we applied our method and evaluated the results for a locus known to contain one of the strongest signals of selection detected so far, and obtained a value for s consistent with current estimates.

It remains an open question if there exists an optimal algorithm for finding all maximal perfect haplotype blocks, *i.e.*, an algorithm that runs in $O(kn)$ time.

It could be worthwhile to study the bipartite graph $(U \cup W, E)$ in which the vertices in $U = \{u_1, \ldots, u_k\}$ correspond to the sequences in S and the vertices in $W = \{w_1, \ldots, w_n\}$ to index positions in these sequences. An edge $(u_i, w_j) \in E$ is drawn if and only if $s_i[j] = 1$. Problem 1 is then equivalent to finding all twin vertices (sets of vertices with identical neighborhood) in intervals of vertices in W. Figure 4 shows this graph for the sequences from Example 2.

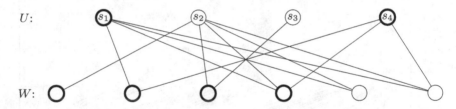

Fig. 4. Bipartite graph $(U \cup W, E)$ representing the four binary sequences $s_1 = 010111$, $s_2 = 101111$, $s_3 = 001000$ and $s_4 = 010101$. Haplotype blocks can be identified as sets of twins when the vertices in the lower row W are restricted to a consecutive interval. For example, s_1 and s_4 are twins in the interval formed by the first four vertices of W (indicated by thick circles), corresponding to the maximal perfect haplotype block 0101.

Twin vertices of a graph G can be determined by constructing its modular decomposition tree, where internal nodes labeled as *series* or *parallel* correspond to last descendant leaves which are twin vertices in G. McConnell and Montgolfier [7] proposed an algorithm to build a modular decomposition tree of a graph with $|V|$ vertices and $|E|$ edges that runs in $O(|V| + |E|)$ time. Since there are $O(n^2)$ necessary subgraphs to detect twin vertices, so far, such strategy is not better than the one we proposed in Algorithm 1. However, it might be possible to achieve some improvement using the fact that intervals in W are not independent.

Another alternative approach could be to use a generalized suffix tree of all the input sequences or the positional Burrows Wheeler Transform [5,8].

References

1. 1000 Genomes Project Consortium, Auton, A., et al.: A global reference for human geneticvariation. Nature **526**(7571), 68–74 (2015)
2. Browning, S.R., Browning, B.L.: Rapid and accurate haplotype phasing and missing-data inference for whole-genome association studies by use of localized haplotype clustering. Am. J. Hum. Genet. **81**(5), 1084–1097 (2007)
3. Chen, H., Hey, J., Slatkin, M.: A hidden Markov model for investigating recent positive selection through haplotype structure. Theor. Popul. Biol. **99**, 18–30 (2015)
4. Chen, H., Slatkin, M.: Inferring selection intensity and allele age from multi-locus haplotype structure. G3: Genes Genomes, Genet. **3**(8), 1429–1442 (2013)
5. Durbin, R.M.: Efficient haplotype matching and storage using the positional burrows-wheeler transform (PBWT). Bioinformatics **30**(9), 1266–1272 (2014)
6. Hayden, E.C.: Technology: The $1,000 genome. Nature **507**(7492), 294–295 (2014)
7. McConnell, R.M., De Montgolfier, F.: Linear-time modular decomposition of directed graphs. Discret. Appl. Math. **145**(2), 198–209 (2005)
8. Norri, T., Cazaux, B., Kosolobov, D., Mäkinen, V.: Minimum Segmentation for Pan-genomic Founder Reconstruction in Linear Time. In: Proceedings of WABI 2018. LIPIcs, vol. 113, pp. 15:1–15:15 (2018)
9. O'Leary, N.A.: Reference sequence (RefSeq) database at NCBI: current status, taxonomic expansion, and functional annotation. Nucl. Acids Res. **44**(D1), D733–D745 (2016)
10. Pedersen, B.S., Quinlan, A.R.: cyvcf2: fast, flexible variantanalysis withPython. Bioinformatics **33**(12), 1867–1869 (2017)
11. Ségurel, L., Bon, C.: On the evolution of lactase persistence in humans. Ann. Rev. Genomics Hum. Genet. **18**, 297–319 (2017)

Sorting by Weighted Reversals
and Transpositions

Andre Rodrigues Oliveira[1]([✉])[iD], Klairton Lima Brito[1][iD], Zanoni Dias[1][iD],
and Ulisses Dias[2][iD]

[1] Institute of Computing, University of Campinas, Campinas, Brazil
{andrero,klairton,zanoni}@ic.unicamp.br
[2] School of Technology, University of Campinas, Limeira, Brazil
ulisses@ft.unicamp.br

Abstract. *Genome rearrangements* are global mutations that change large stretches of DNA sequence throughout genomes. They are rare but accumulate during the evolutionary process leading to organisms with similar genetic material in different places and orientations within the genome. Sorting by Genome Rearrangements problems seek for minimum-length sequences of rearrangements that transform one genome into the other. These problems accept alternative versions that assign weights for each event and the goal is to find a *minimum-weight* sequence. We study the *Sorting by Weighted Reversals and Transpositions* problem in two variants depending on whether we model genomes as signed or unsigned permutations. Here, we use weight 2 for reversals and 3 for transpositions and consider theoretical and practical aspects in our analysis.

We present one algorithm with an approximation factor of 2 for both signed or unsigned permutations, and one algorithm with an approximation factor of $\frac{5}{3}$ for signed permutations. We also analyze the behavior of the $\frac{5}{3}$-approximation algorithm with different weights for reversals and transpositions.

Keywords: Genome rearrangement · Weighted operations
Reversals and transpositions · Approximation algorithms

1 Introduction

A *genome rearrangement* is a mutational event that affects large portions of a genome. The rearrangement distance is the minimum number of events that transform one genome into another and serves as approximation for the evolutionary distance due to the *principle of parsimony*.

We define genomes as permutations of integer numbers where each number is a gene or a set of genes. These permutations can be *signed* or *unsigned*, depending on which information we have about gene orientation. To compute the rearrangement distance between two genomes, we represent one of them as

© Springer Nature Switzerland AG 2018
R. Alves (Ed.): BSB 2018, LNBI 11228, pp. 38–49, 2018.
https://doi.org/10.1007/978-3-030-01722-4_4

the identity permutation (where all elements are sorted) to have a *Sorting by Genome Rearrangements* problem.

A *model* $\mathcal{M}$ is a set of rearrangements allowed to act on a genome. Reversals and transpositions are the best known genome rearrangements. A reversal occurs when, given two positions of a genome, the sequence between these positions is reversed and, on signed permutations, all signs between these positions are flipped. Transpositions occur when two consecutive sequences inside a genome exchange position.

When the model allows only reversals we have the *Sorting by Reversals* (**SbR**) problem. On signed permutations, **SbR** has a polynomial algorithm [14]. On unsigned permutations, **SbR** is an NP-hard Problem [7], and the best approximation factor is 1.375 [4].

When the model allows only transpositions we have the *Sorting by Transpositions* (**SbT**) problem. The **SbT** is an NP-hard problem [6], and the best approximation factor is 1.375 [10].

When the model allows both reversals and transpositions we have the *Sorting by Reversals and Transpositions* (**SbRT**) problem. The complexity of this problem is unknown on signed and unsigned permutations. On signed permutations, the best approximation factor is 2 [17]. On unsigned permutations, the best approximation factor is $2k$ [16], where k is the approximation factor of an algorithm used for cycle decomposition [7]. Given the best value for k known so far [8], this approximation factor is $2.8334 + \epsilon$ for some $\epsilon > 0$.

A variant of problems considers different weights for different rearrangements, which is useful if a specific rearrangement is more likely to occur than others in a particular population [5,18]. Gu *et al.* [13] introduced the rearrangement event called *inverted transposition*, a transposition where one of the blocks is reversed, and Eriksen [12] developed an algorithm with an approximation factor of 7/6 and a polynomial-time approximation scheme (PTAS) when reversals have weight 1 and transpositions and inverted transpositions have weight 2. Later, Bader and Ohlebusch [2] developed an algorithm with approximation factor of 1.5 when reversals have weight 1 and transpositions and inverted transpositions have both same weight in the range [1..2].

We present two approximation algorithms for the *Sorting by Weighted Reversals and Transpositions* problem (**SbWRT**) such that the weight of a reversal (w_ρ) is 2 and the weight of a transposition (w_τ) is 3. These weights were considered optimal by Eriksen [11] on his experiments. Bader and coauthors [1] also developed experiments adopting exactly this weight ratio, and they claim that it equilibrates the use of reversals and transpositions and is realistic for most biological datasets.

The paper is organized as follows. Section 2 presents the background we use throughout the paper. Section 3 presents two approximation algorithms for **SbWRT**, with approximation factors of 2 and 5/3. Section 3 also investigates other weights in the 5/3-approximation algorithm. Section 4 concludes the paper.

2   Background

We model a genome with n genes as an n-tuple whose elements represent genes. Assuming no duplicated gene, this n-tuple is as a permutation $\pi = (\pi_1\ \pi_2\ ... \pi_{n-1}\ \pi_n)$ with $|\pi_i| \in \{1, 2, ..., (n-1), n\}$, for $1 \le i \le n$, and $|\pi_i| = |\pi_j|$ if, and only if, $i = j$. If we know the relative orientation of the genes, we represent it with signs $+$ or $-$ assigned to each element, and we say that π is a *signed* permutation. Otherwise, the signs are omitted and π is an *unsigned* permutation.

The *identity permutation* is the permutation in which all elements are in ascending order and, in signed permutations, have positive sign, and we denote this permutation by ι.

An *extended permutation* can be obtained from a permutation π by adding elements $\pi_0 = 0$ and $\pi_{n+1} = n+1$ in the unsigned case, or $\pi_0 = +0$ and $\pi_{n+1} = +(n+1)$ in the signed case. From now on, unless stated otherwise, we will refer to an extended permutation as "permutation" only.

A *reversal* $\rho(i, j)$ with $1 \le i \le j \le n$ reverts the order of the segment $\{\pi_i, \pi_{i+1}, ..., \pi_j\}$. When π is a signed permutation, the reversal $\rho(i, j)$ also flips the signs of the elements in $\{\pi_i, \pi_{i+1}, ..., \pi_j\}$. Therefore, if π is a signed permutation then $(\pi_1\ ...\ \pi_i\ ...\ \pi_j\ ...\ \pi_n) \circ \rho(i,j) = (\pi_1\ ...\ \underline{-\pi_j\ ...\ -\pi_i}\ ...\ \pi_n)$ and if π is an unsigned permutation we have that $(\pi_1\ ...\ \pi_i\ ...\ \pi_j\ ...\ \pi_n) \circ \rho(i,j) = (\pi_1\ ...\ \underline{\pi_j\ ...\ \pi_i}\ ...\ \pi_n)$. The *weight of a reversal* $\rho(i, j)$ is denoted by w_ρ.

A *transposition* $\tau(i, j, k)$ with $1 \le i < j < k \le n+1$ exchanges the blocks $\{\pi_i, \pi_{i+1}, ..., \pi_{j-1}\}$ and $\{\pi_j, \pi_{j+1}, ..., \pi_{k-1}\}$. Since these blocks are not reversed, transpositions never change signs. Therefore, given a signed or unsigned permutation π, we have that $(\pi_1\ ...\ \pi_i\ \underline{\pi_i\ ...\ \pi_{j-1}}\ \underline{\pi_j\ ...\ \pi_{k-1}}\ \pi_k\ ...\ \pi_n) \circ \tau(i,j,k) = (\pi_1\ ...\ \pi_i\ \underline{\pi_j\ ...\ \pi_{k-1}}\ \underline{\pi_i\ ...\ \pi_{j-1}}\ \pi_k\ ...\ \pi_n)$. We denote by w_τ the *weight of a transposition* $\tau(i, j, k)$.

Let $\mathcal{S}$ be a sequence of reversals and transpositions that sorts π, we denote the cost of $\mathcal{S}$ by $w_\mathcal{S}(\pi, w_\rho, w_\tau) = w_\rho \times S_\rho + w_\tau \times S_\tau$ such that S_ρ and S_τ are the number of reversals and transpositions in $\mathcal{S}$, respectively.

We denote the *weighted distance* of π by $w_{RT}(\pi, w_\rho, w_\tau) = w_{\mathcal{S}'}(\pi, w_\rho, w_\tau)$ such that $\mathcal{S}'$ is a sequence of operations that sorts π and for any sequence $\mathcal{S}$ that sorts π, $w_{\mathcal{S}'}(\pi, w_\rho, w_\tau) \le w_\mathcal{S}(\pi, w_\rho, w_\tau)$. Since we stated that $w_\rho = 2$ and $w_\tau = 3$, given a sequence $\mathcal{S}$ with S_ρ reversals and S_τ transpositions we have that $w_\mathcal{S}(\pi, w_\rho = 2, w_\tau = 3) = 2S_\rho + 3S_\tau$.

2.1   Breakpoints and Strips

Let π be an unsigned permutation, a pair of consecutive elements (π_i, π_{i+1}), with $0 \le i \le n$, is a *breakpoint* if $|\pi_{i+1} - \pi_i| \ne 1$. Let π be a signed permutation, a pair of consecutive elements (π_i, π_{i+1}), with $0 \le i \le n$, is a *breakpoint* if $\pi_{i+1} - \pi_i \ne 1$. The total number of breakpoints in both cases is $b(\pi)$. The identity permutation is the only permutation without breakpoints ($b(\iota) = 0$).

A reversal $\rho(i, j)$ acts on two points of π, so given a permutation π and a reversal ρ, we have $b(\pi) - 2 \le b(\pi \circ \rho) \le b(\pi) + 2$. A transposition $\tau(i, j, k)$ acts on three points of π, so given a permutation π and a transposition τ, we have $b(\pi) - 3 \leq b(\pi \circ \tau) \leq b(\pi) + 3$.

Let $\Delta b(\pi, S) = b(\pi) - b(\pi \circ S)$ denote the variation in the number of breakpoints by applying a sequence S of operations to π. With $w_\rho = 2$ and $w_\tau = 3$ we have that $\Delta b(\pi, \rho) \leq w_\rho$ and $\Delta b(\pi, \tau) \leq w_\tau$, so given any sequence S of operations, $\Delta b(\pi, S) \leq w_S(\pi, w_\rho = 2, w_\tau = 3)$, which results in the following lemma.

Lemma 1. *Given a sequence S and a permutation π, if $\Delta b(\pi, S) > 0$ then the approximation factor of S is $w_S(\pi, w_\rho = 2, w_\tau = 3)/\Delta b(\pi, S)$.*

Breakpoints divide the permutation into *strips*, which are maximal subsequences of consecutive elements with no breakpoint. The first strip of a permutation always starts with element π_1 and the last strip always ends with element π_n (the elements π_0 and π_{n+1} are not computed as strips). On signed permutations, a strip is *positive* if its elements have positive signs, and it is *negative* otherwise. On unsigned permutations, a strip is *positive* if the displacement of its elements from left to right forms an increasing sequence, and it is *negative* otherwise. By convention, strips with only one element are negative on unsigned permutations.

For instance, the permutation $\pi = (+0\ +2\ +3\ -1\ -6\ -5\ +4\ +7)$ has $b(\pi) = 5$ since the pairs $(+0, +2)$, $(+3, -1)$, $(-1, -6)$, $(-5, +4)$, and $(+4, +7)$ are breakpoints. Besides, π has four strips: the positive strips $(+2, +3)$ and $(+4)$ and the negative strips (-1) and $(-6, -5)$.

2.2 Cycle Graph

Cycle graph is a tool to develop non trivial bounds for sorting problems using reversals and/or transpositions [3]. Given a signed permutation π, we create its cycle graph $G(\pi)$ as follows. For each element π_i, with $1 \leq i \leq n$, we add to the set $V(G(\pi))$ two vertices: $-\pi_i$ and $+\pi_i$. Finally we add the vertices $+\pi_0$ and $-\pi_{n+1}$. The set of edges $E(G(\pi))$ is formed by two types of edges: black edges and gray edges. The set of black edges is $\{(-\pi_i, +\pi_{i-1}) \mid 1 \leq i \leq n+1\}$. The set of gray edges is $\{(+(i-1), -i) \mid 1 \leq i \leq n+1\}$. Since each vertex is incident to one gray edge and one black edge, there is a unique decomposition of edges in cycles. Note that $G(\iota)$ has $(n+1)$ trivial cycles (i.e., cycles with one black edge and one gray edge).

When π is unsigned, we create a signed permutation π' by randomly assigning signs to its elements, then we construct the cycle graph $G(\pi')$ and associate $G(\pi')$ to π. Therefore, an unsigned permutation π have an exponential number of cycle graph representations. Since our goal is to transform π into ι, the best cycle graph maximizes the number of cycles. Finding this cycle graph for unsigned permutations is an NP-hard problem [7].

The *size* of a cycle $c \in G(\pi)$ is the number of black edges in c. A cycle c is *odd* if its size is odd, and it is *even* otherwise. A cycle is *short* if it contains three or less black edges, and it is *long* otherwise. We denote by $c(G(\pi))$ the number of cycles in $G(\pi)$ and by $c_{odd}(G(\pi))$ the number of odd cycles in $G(\pi)$.

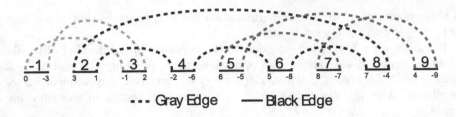

--- Gray Edge   —— Black Edge

Fig. 1. Cycle graph of permutation $\pi = (+3\ -1\ -2\ +6\ +5\ +8\ +7\ +4)$.

We draw the cycle graph $G(\pi)$ in a way that reveals important features (Fig. 1). For each element $\pi_i \in \pi$, with $1 \leq i \leq n$, we place the vertex $-\pi_i$ before the vertex π_i. The vertex $+\pi_0$ is located in the left of $-\pi_1$ and the vertex $-\pi_{i+1}$ is located in the right of π_n.

Black edges of $G(\pi)$ are labeled from 1 to $n + 1$ so that the black edge $(-\pi_i, \pi_{i-1})$ is labeled as i. We represent c as the sequence of labels of its black edges following the order that they appear in the path starting from the rightmost black edge being traversed from right to left.

If a black edge i is traversed from left to right we label it as $-i$; otherwise, we simply label it as i.

Two black edges in a cycle c are *convergent* if they have the same sign, and it is *divergent* otherwise. A cycle c is *divergent* if it has two divergent black edges, and it is *convergent* otherwise. Moreover, a convergent cycle $c = (i_1, ..., i_k)$, for all $k > 1$, is *nonoriented* if $i_1, ..., i_k$ is a decreasing sequence, and c is *oriented* otherwise. Two cycles $c_1 = (i_1, ..., i_k)$ and $c_2 = (j_1, ..., j_k)$ with $k > 1$ are *interleaving* if $|i_1| > |j_1| > |i_2| > |j_2| > ... > |i_k| > |j_k|$ or $|j_1| > |i_1| > |j_2| > |i_2| > ... > |j_k| > |i_k|$.

Let g_1 and g_2 be two gray edges such that: (i) g_1 is incident to black edges labeled as x_1 and y_1, (ii) g_2 is incident to black edges labeled as x_2 and y_2, (iii) $|x_1| < |y_1|$, and (iv) $|x_2| < |y_2|$. We say that g_1 *intersects* g_2 if $|x_1| < |x_2| < |y_1| < |y_2|$ or $|x_1| < |x_2| = |y_1| < |y_2|$. Two cycles c_1 and c_2 intersects if a gray edge $g_1 \in c_1$ intersects a gray edge $g_2 \in c_2$.

Figure 1 shows the cycle graph for $\pi = (+3\ -1\ -2\ +6\ +5\ +8\ +7\ +4)$. There are three cycles in $G(\pi)$: the oriented odd cycle $c_1 = (+9, +5, +7)$; the nonoriented even cycle $c_2 = (+8, +6, +4, +2)$; and the divergent even cycle $c_3 = (+3, -1)$ (the black edge 1 has minus sign because it is traversed from left to right). Moreover, gray edge $g_1 \in c_3$, which links black edges 1 and 3, intersects with gray edge $g_2 \in c_2$, which links black edges 2 and 4, so it follows that cycles c_2 and c_3 intersects.

A permutation π is *simple* if $G(\pi)$ contains only short cycles. We can transform any permutation π into a simple permutation $\hat{\pi}$ by adding new elements to break long cycles. This is a common approach for sorting permutations by reversals [14] and by transpositions [10]. Let c be a cycle of size $k \geq 4$, we add a black edge (i.e., a new element in the permutation) to transform c into two cycles c_1 with 3 black edges and c_2 with $k - 2$ black edges, as shown in Fig. 2.

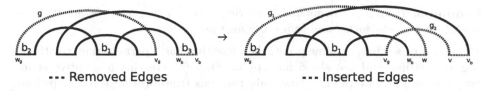

··· Removed Edges ··· Inserted Edges

Fig. 2. Transformation of a long cycle of size $k = 5$ into two cycles of size 3. The transformation is made by removing edges g and b_3, adding two vertices w and b between w_b and v_b, adding two gray edges $g_1 = (w_g, w)$ and $g_2 = (v_g, v)$ and two black edges (w_b, w) and (v, v_b).

3 Approximation Algorithms for SbWRT

We develop two approximation algorithms: one for signed or unsigned permutations with approximation factor of 2, and one specific for signed permutations with an approximation factor of 5/3.

3.1 The 2-Approximation Algorithm

Our first approximation algorithm uses breakpoints and strips to sort signed or unsigned permutations. It is a greedy algorithm that removes as many breakpoints as possible at each iteration and tries to keep negative strips. At each step, while the permutation is not sorted, the algorithm searches for an operation using the following order of priority:

(i) a transposition that removes three breakpoints;
(ii) a reversal that removes two breakpoints;
(iii) a transposition that removes two breakpoints;
(vi) a reversal that removes one breakpoint and keeps a negative strip;
(v) a reversal applied to the first two breakpoints of the permutation.

Note that Step (iv), which removes one breakpoint with a reversal, is applied only if it keeps at least one negative strip. Step (v) is the only that does not remove any breakpoint, but it creates a negative strip. According to Lemma 1, Steps (i) and (ii) have an approximation factor of 1, Step (iii) has an approximation factor of 3/2, and Step (iv) has an approximation of 2. Since Step (v) does not remove breakpoints, we first recall the following lemma to ensure the approximation factor 2 for this algorithm.

Lemma 2. [15] *Let π be a permutation with a negative strip. If every reversal that removes a breakpoint of π leaves a permutation with no negative strips, π has a reversal that removes two breakpoints.*

The following lemma guarantees that every time the algorithm perform Step (v), it performs Step (ii) before reaching the identity permutation and before Step (v) is performed again.

Lemma 3. *If Step (v) is applied, then Step (ii) will be applied before the permutation becomes sorted and before the next use of Step (v).*

Proof. Suppose that, at some point, the algorithm has a permutation π without negative strips, and the algorithm applies Step (v) creating a negative strip. By Lemma 2 and by the fact that only reversals transform negative strips into positive ones, we know that the algorithm will perform Step (ii). But before performing Step (ii) the permutation will always have at least one negative strip, and Step (v) will not be performed until then. □

From Lemma 3, each occurrence of Step (v) forces a later occurrence of Step (ii), which gives us, by Lemma 1, the approximation factor of 2. We can find which step of the algorithm can be applied in linear time, and this process repeats at most n times. Thus, in the worst case, the algorithm runs in $O(n^2)$.

3.2 The 5/3-Approximation Algorithm

This algorithm uses cycle graph to sort a permutation. We define a score function to prove an approximation factor of $\frac{5}{3}$ for signed permutations based on the number of cycles and the number of odd cycles.

Given a sequence $\mathcal{S}$ of operations applied on π, we denote by $\Delta c(\pi, \mathcal{S}) = c(G(\pi \circ \mathcal{S})) - c(G(\pi))$ and $\Delta c_{odd}(\pi, \mathcal{S}) = c_{odd}(G(\pi \circ \mathcal{S})) - c_{odd}(G(\pi))$ the variation in the number of cycles and odd cycles, respectively, caused on $G(\pi)$ by $\mathcal{S}$.

Let w_c be the weight of a cycle in $G(\pi)$ and w_o be the weight of an odd cycle in $G(\pi)$. Using these weights, we define a score function that relates the gain in terms of cycles and odd cycles after applying a sequence $\mathcal{S}$ of operations and the cost of operations in $\mathcal{S}$. The cost of a sequence $\mathcal{S}$ is $w_{\mathcal{S}}(\pi, w_\rho, w_\tau)$, and $G(\pi \circ \mathcal{S})$ has a variation of $\Delta c(\pi, \mathcal{S})$ cycles and $\Delta c_{odd}(\pi, \mathcal{S})$ odd cycles compared to $G(\pi)$.

The *objective function*, denoted by $R_{\mathcal{S}}$, is then defined as

$$R_{\mathcal{S}} = \frac{w_c \Delta c(\pi, \mathcal{S}) + w_o \Delta c_{odd}(\pi, \mathcal{S})}{w_{\mathcal{S}}(\pi, w_\rho, w_\tau)}.$$

Note that $\Delta c(\pi, \rho) \in \{-1, 0, 1\}$ and $\Delta c_{odd}(\pi, \rho) \in \{-2, 0, 2\}$ [9,14], so let $R_\rho = \frac{w_c + 2w_o}{2}$ be the best objective function of a reversal. Similarly, $\Delta c(\pi, \tau) \in \{-2, -1, 0, 1, 2\}$ and $\Delta c_{odd}(\pi, \tau) \in \{-2, 0, 2\}$ [3,9,17], so let $R_\tau = \frac{2w_c + 2w_o}{3}$ be the best objective function of a transposition. We have the following lower bound regarding the number of cycles and odd cycles.

Lemma 4. *Given a permutation π we have that*

$$w_{RT}(\pi, w_\rho, w_\tau) \geq \frac{(w_c + w_o)(n+1) - (w_c \times c(\pi) + w_o \times c_{odd}(\pi))}{\max\{R_\rho, R_\tau\}}$$

Proof. Note that for any reversal $\rho(i,j)$ we have $R_{\rho(i,j)} \leq R_\rho$, and for any transposition $\tau(i,j,k)$ we have $R_{\tau(i,j,k)} \leq R_\tau$. Since $G(\iota)$ has $n+1$ odd cycles, a lower bound can be obtained by dividing the necessary gain in terms of cycles and odd cycles to reach the identity permutation (in this case $(w_c + w_o)(n+1) - (w_c \times c(\pi) + w_o \times c_{odd}(\pi)))$ by the best ratio function between reversals and transpositions $(\max\{R_\rho, R_\tau\})$. □

The procedure to transform π into a simple permutation $\hat{\pi}$ (Sect. 2.2) docs not guarantee that $w_{RT}(\hat{\pi}, w_\rho, w_\tau) = w_{RT}(\pi, w_\rho, w_\tau)$, but it guarantees π and $\hat{\pi}$ have the same lower bound given by Lemma 4, since it adds a new cycle to the identity permutation (so, instead $(n+1)$, we have $(n+2)$ in the first part of the dividend), and it also adds a new odd cycle to the permutation (so both $c(\pi)$ and $c_{odd}(\pi)$ increases by one in the second part of the dividend). Therefore, our algorithm starts by transforming π into a simple permutation $\hat{\pi}$.

Until the permutation is sorted, the algorithm searches for an operation using the following order of priority:

(i) If there is an oriented cycle C, apply a transposition on C (Fig. 3(i)).
(ii) If there is a divergent cycle C of size two, apply a reversal on C (Fig. 3(ii)).
(iii) If there is a divergent cycle C of size three, apply a reversal on two divergent black edges of C (Fig. 3(iii)).
(vi) If there are two intersected cycles C_1 and C_2 of size two, apply a sequence of two transpositions on C_1 and C_2 (Fig. 3(iv)).
(v) If there is a cycle C_1 of size 3 intersected by two cycles C_2 and C_3 both of size two, apply a sequence of two transpositions on these cycles (Fig. 3(v)).
(vi) If there is a non-oriented cycle C_1 of size 3 interleaved with a non-oriented cycle C_2 of size 3, apply a sequence of three transpositions on these cycles (Fig. 3(vi)).
(vii) If there is a cycle C_1 of size 3 intersected by two cycles C_2 and C_3 both of size three, apply a sequence of three transpositions on these cycles (Fig. 3(vii)).
(viii) If there is a cycle C_1 of size 3 intersected by two cycles C_2 of size two and C_3 of size three, apply a sequence of three transpositions on these cycles (Fig. 3(vii)).

Lemma 5. *If a permutation π is such that $\pi \neq \iota$ then it is always possible to perform at least one of the eight steps above.*

Proof. Since $\pi \neq \iota$, $G(\pi)$ cannot contain only trivial cycles. Let $\hat{\pi}$ be the simple permutation obtained by the procedure that breaks long cycles from $G(\pi)$.

If $G(\hat{\pi})$ has oriented or divergent cycles, Steps (i) to (iii) can be applied. Otherwise, $G(\hat{\pi})$ has only trivial and nonoriented cycles of size two or three. Bafna and Pevzner [3] showed that each nonoriented cycle of size two must intersect with at least another cycle, and each nonoriented cycle of size three must either interleave with another cycle, or intersect with at least two nonoriented cycles.

If $G(\hat{\pi})$ has a nonoriented cycle of size three, this cycle either interleaves with another nonoriented cycle (Step (vi) can be applied), or it intersects with two other cycles, and these two cycles are: both even (Step (v) can be applied), or both odd (Step (vii) can be applied), or have different parities (Step $(viii)$ can be applied). Otherwise, $G(\hat{\pi})$ has a nonoriented cycle of size two that intersects with another cycle of size two (Step (iv) can be applied). □

Figure 3 illustrates each step and shows the cycles generated. Step (iii) generates one trivial cycle and one cycle of size two, that can be divergent or not

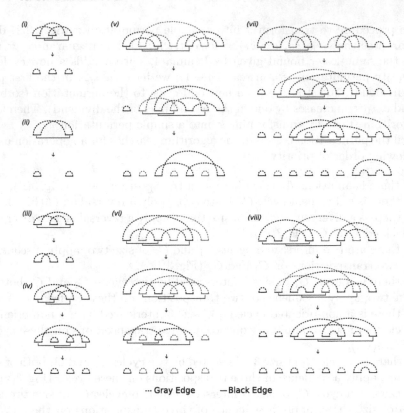

Fig. 3. Schema of operations applied and cycle representation in each of the eight steps of the algorithm.

depending on the input cycle. Step (v) generates four trivial cycles and one cycle of size three, that can be oriented or not depending on the input two cycles.

Step (vii) generates six trivial cycles and one cycle of size three, that is nonoriented. Step $(viii)$ generates six trivial cycles and one cycle of size two, that is nonoriented. All the remaining steps generate only trivial cycles. Note that all steps keep the cycle graph simple since they generate cycles with three or less black edges.

Table 1 summarizes for each step the variation on cycles ($\Delta c(\pi, \mathcal{S})$), the variation on odd cycles ($\Delta c_{odd}(\pi, \mathcal{S})$), the sequence applied ($\mathcal{S}$), the weight of the sequence applied ($w_{\mathcal{S}}(\pi, w_\rho, w_\tau)$), and its objective function ($R_{\mathcal{S}}$).

The best approximation factor in this case is achieved when $w_c = 4$ and $w_o = 1$, resulting in the following values of objective functions: $R_{(i)} = \frac{10}{3}$; $R_{(ii)} = \frac{6}{2} = 3$; $R_{(iii)} = \frac{4}{2} = 2$; $R_{(iv)}, R_{(v)} = \frac{12}{6} = 2$; $R_{(vi)}, R_{(vii)}, R_{(viii)} = \frac{20}{9}$. The greatest objective function is $\frac{10}{3}$ and the lowest is 2, which give us the approximation factor of $\frac{\frac{10}{3}}{2} = \frac{10}{6} = \frac{5}{3} \approx 1.667$. We can find which step of the algorithm can be applied in quadratic time, and this process repeats at most n times. Thus, in the worst case, the algorithm runs in $O(n^3)$.

Table 1. Summary of variation on number of cycles and the objective function of each step performed by the algorithm.

Step	S	$w_S(\pi, w_\rho, w_\tau)$	$\Delta c(\pi, S)$	$\Delta c_{odd}(\pi, S)$	R_S
(i)	τ	3	2	2	$\frac{2w_c + 2w_o}{3}$
(ii)	ρ	2	1	2	$\frac{1w_c + 2w_o}{2}$
(iii)	ρ	2	1	0	$\frac{1w_c}{2}$
(iv)	τ, τ	6	2	4	$\frac{2w_c + 4w_o}{6}$
(v)	τ, τ	6	2	4	$\frac{2w_c + 4w_o}{6}$
(vi)	τ, τ, τ	9	4	4	$\frac{4w_c + 4w_o}{9}$
(vii)	τ, τ, τ	9	4	4	$\frac{4w_c + 4w_o}{9}$
(viii)	τ, τ, τ	9	4	4	$\frac{4w_c + 4w_o}{9}$

3.3   Investigating Different Weights for Reversals and Transpositions

We investigate different values for w_ρ and w_τ in our 5/3-approximation algorithm. We generated alternative sequences that produce the same cycle graph as the original sequences for each step of the algorithm, except for steps (ii) and (iii) that apply only one reversal each. Table 2 lists the alternative sequences.

Table 2. Alternative sequences to the 5/3-approximation algorithm.

Step	Original sequence	Alternative sequence(s)
(i)	τ	ρ, ρ, ρ
(iv)	τ, τ	ρ, ρ, ρ
(v)	τ, τ	ρ, ρ, ρ, ρ
(vi)	τ, τ, τ	$\rho, \rho, \rho, \rho, \rho$ and $\tau, \tau, \rho, \rho, \rho$
(vii)	τ, τ, τ	$\tau, \tau, \rho, \rho, \rho$ and $\rho, \rho, \rho, \rho, \rho, \rho$ and $\tau, \rho, \rho, \rho, \rho, \rho$
(viii)	τ, τ, τ	$\rho, \rho, \rho, \rho, \rho$ and $\tau, \tau, \rho, \rho, \rho$

Any transposition $\tau(i, j, k)$ can be replaced by three reversals $\rho(i, j-1)\rho(j, k-1)\rho(i, k-1)$, so we stand as a limit that $w_\tau \leq 3w_\rho$, otherwise the algorithm can apply only reversals. For each pair of w_ρ and w_τ, we tested different values for w_c and w_o. For each value of w_τ/w_ρ we calculated the maximum value of objective function between the original and the alternative sequences, and obtained the approximation factor using the greatest and lowest objective functions. In our analysis, the approximation factor is always less than two when w_τ/w_ρ is in the interval $1 < w_\tau/w_\rho < 2$.

As shown in Fig. 4, it is possible to improve the approximation factor to $89/55 \approx 1.618$ by allowing the alternative sequences from Table 2, and using values of w_τ and w_ρ such that $w_\tau/w_\rho = 1.618$, $w_c = 55$, and $w_o = 17$.

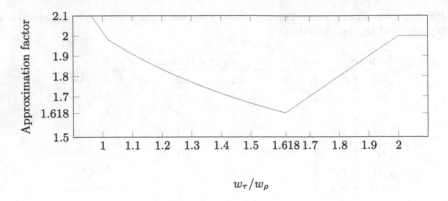

Fig. 4. Approximation factor for different values of w_ρ and w_τ.

4   Conclusion

We studied the problem of Sorting by Weighted Reversals and Transpositions when the weight of a reversal is 2 and the weight of a transposition is 3. We developed two different algorithms: a simpler 2-approximation algorithm for both signed and unsigned permutations using breakpoints and strips, and a more elaborated $5/3 \approx 1.667$-approximation algorithm for signed permutations using cycle graphs.

We also investigated different values for the weight of a reversal and a transposition, and came to the conclusion that we can obtain a 1.618-approximation algorithm by allowing some alternative sequences at steps of the 1.667-approximation algorithm, and setting $w_\tau/w_\rho = 1.618$, $w_c = 55$, and $w_o = 17$. We intend to continue working with this problem in order to develop algorithms with approximation factors strictly less than 1.6.

Acknowledgments. This work was supported by the National Counsel of Technological and Scientific Development, CNPq (grants 400487/2016-0, 425340/2016-3, and 140466/2018-5), the São Paulo Research Foundation, FAPESP (grants 2013/08293-7, 2015/ 11937-9, 2017/12646-3, 2017/16246-0, and 2017/16871-1), the Brazilian Federal Agency for the Support and Evaluation of Graduate Education, CAPES, and the CAPES/COFECUB program (grant 831/15).

References

1. Bader, M., Abouelhoda, M.I., Ohlebusch, E.: A fast algorithm for the multiple genome rearrangement problem with weighted reversals and transpositions. BMC Bioinform. **9**(1), 1–13 (2008)
2. Bader, M., Ohlebusch, E.: Sorting by weighted reversals, transpositions, and inverted transpositions. J. Comput. Biol. **14**(5), 615–636 (2007)
3. Bafna, V., Pevzner, P.A.: Sorting by transpositions. SIAM J. Discret. Math. **11**(2), 224–240 (1998)

4. Berman, P., Hannenhalli, S., Karpinski, M.: 1.375-approximation algorithm for sorting by reversals. In: Möhring, R., Raman, R. (eds.) ESA 2002. LNCS, vol. 2461, pp. 200–210. Springer, Heidelberg (2002). https://doi.org/10.1007/3-540-45749-6_21

5. Blanchette, M., Kunisawa, T., Sankoff, D.: Parametric genome rearrangement. Gene **172**(1), GC11–GC17 (1996)

6. Bulteau, L., Fertin, G., Rusu, I.: Sorting by transpositions is difficult. SIAM J. Discret. Math. **26**(3), 1148–1180 (2012)

7. Caprara, A.: Sorting permutations by reversals and eulerian cycle decompositions. SIAM J. Discret. Math. **12**(1), 91–110 (1999)

8. Chen, X.: On sorting unsigned permutations by double-cut-and-joins. J. Comb. Optim. **25**(3), 339–351 (2013)

9. Dias, U., Galvão, G.R., Lintzmayer, C.N., Dias, Z.: A general heuristic for genome rearrangement problems. J. Bioinform. Comput. Biol. **12**(3), 26 (2014)

10. Elias, I., Hartman, T.: A 1.375-approximation algorithm for sorting by transpositions. IEEE/ACM Trans. Comput. Biol. Bioinform. **3**(4), 369–379 (2006)

11. Eriksen, N.: Combinatorics of Genome Rearrangements and Phylogeny. Teknologie licentiat thesis, Kungliga Tekniska Högskolan, Stockholm (2001)

12. Eriksen, N.: (1+ε)-approximation of sorting by reversals and transpositions. Theor. Comput. Sci. **289**(1), 517–529 (2002)

13. Gu, Q.P., Peng, S., Sudborough, I.H.: A 2-approximation algorithm for genome rearrangements by reversals and transpositions. Theor. Comput. Sci. **210**(2), 327–339 (1999)

14. Hannenhalli, S., Pevzner, P.A.: Transforming cabbage into turnip: polynomial algorithm for sorting signed permutations by reversals. J. ACM **46**(1), 1–27 (1999)

15. Kececioglu, J.D., Sankoff, D.: Exact and approximation algorithms for sorting by reversals, with application to genome rearrangement. Algorithmica **13**, 180–210 (1995)

16. Rahman, A., Shatabda, S., Hasan, M.: An approximation algorithm for sorting by reversals and transpositions. J. Discret. Algorithms **6**(3), 449–457 (2008)

17. Walter, M.E.M.T., Dias, Z., Meidanis, J.: Reversal and transposition distance of linear chromosomes. In: Proceedings of the 5th International Symposium on String Processing and Information Retrieval (SPIRE 1998), pp. 96–102. IEEE Computer Society, Los Alamitos (1998)

18. Yancopoulos, S., Attie, O., Friedberg, R.: Efficient sorting of genomic permutations by translocation, inversion and block interchange. Bioinformatics **21**(16), 3340–3346 (2005)

Graph Databases in Molecular Biology

Waldeyr M. C. da Silva[1,2](✉) (iD), Polyane Wercelens[2],
Maria Emília M. T. Walter[2], Maristela Holanda[2], and Marcelo Brígido[2]

[1] Federal Institute of Goiás, Formosa, Brazil
waldeyr.mendes@ifg.edu.br
[2] University of Brasília, Brasília, Brazil

Abstract. In recent years, the increase in the amount of data generated in basic social practices and specifically in all fields of research
has boosted the rise of new database models, many of which have been
employed in the field of Molecular Biology. NoSQL graph databases have
been used in many types of research with biological data, especially in
cases where data integration is a determining factor. For the most part,
they are used to represent relationships between data along two main
lines: (i) to infer knowledge from existing relationships; (ii) to represent
relationships from a previous data knowledge. In this work, a short history in a timeline of events introduces the mutual evolution of databases
and Molecular Biology. We present how graph databases have been used
in Molecular Biology research using High Throughput Sequencing data,
and discuss their role and the open field of research in this area.

Keywords: Graph databases · Molecular Biology · Omics
Contributions

1 Introduction

The development of Molecular Biology precedes the development of the modern Turing-machine based Computer Sciences. However, from the moment they
meet up, this close and cooperative relationship has intensified and accelerated
advances in the field. In 1953, the structure of DNA was described by Watson
and Crick [39] paving the way for the central dogma of Molecular Biology, which
has been continuously enhanced by the discoveries of science. In the same year,
International Business Machines (IBM) launched its first digital computer, the
IBM 701. In the decade that followed, the genetic code was deciphered with the
triplet codon pattern identification [12], which was almost wholly cracked in the
following years [21]. Meanwhile, the history of modern databases began in 1964
at General Electric, when the first considered commercial Database Management
System (DBMS) was born and named IDS - Integrated Data Store [3], [2]. From
then on, other initiatives appeared, such as Multivalue [15], MUMPS [25], and
IMS [13], which was designed for the Apollo space program in 1966.

The 1970s brought in the relational model [8], a very significant database
model that, according to the site DB-Engines, even today, it is the most widely

R. Alves (Ed.): BSB 2018, LNBI 11228, pp. 50–57, 2018.
https://doi.org/10.1007/978-3-030-01722-4_5

used throughout the world. The first DNA sequencing was completed in 1971 [41], and in 1975, Sanger's method [28] for DNA sequencing, led to a generation of methods [18]. Due to the possibility of using computers to analyze DNA sequence data, in 1976 the first COBOL program to perform this type of analysis was published [22] and could be considered the birth of Bioinformatics, even though the name had yet to be coined.

In the 1980s, discussions about the human genome naturally emerged with the advances in DNA sequencing which were due to the affordable costs [29]. Throughout the 1990s, the Human Genome Project conducted the sequencing of the human DNA, which in 2001, culminated in the publications of the two competitors in this assignment [19,38]. Also, in the 1990s, the modern Internet emerged, and in the second half of the 1990s, the world experienced the Internet bubble phenomenon [5].

The efforts of the genome projects have promoted new technologies for sequencing, such as the High-Throughput Sequencing technologies (HTS), which have been used in laboratories worldwide. Nowadays, biological data has increased intensely in volume and diversity, becoming known as omics (genomics, transcriptomics, epigenomics, proteomics, metabolomics, and others). Projections on omics are impressive, and it is estimated that in 2025, genomics will generate one zetta-bases per year, which enables us to characterize the omics as a Big Data science [32]. NoSQL databases have played a significant role in managing large volumes of data, and as in other areas, the omics recently became a target of the NoSQL movement.

Although the NoSQL movement does not have a consensual definition, the literature points out that NoSQL is an umbrella term for non-relational database systems that provide mechanisms for storing and retrieving data, and which has modeling that is an alternative to traditional relational databases and their Structured Query Language (SQL). According to Corbellini [10], there are different types of NoSQL database models commonly classified as key-value, wide column or column families, document-oriented, and graph databases. Despite this classification, the NoSQL databases may be hybrids, using more than one database model.

In this review, we summarize the current NoSQL graph databases contributions to Molecular Biology, limited to their use in omics data from HTS, from the time they were first reported in the literature. We approach the contributions of NoSQL graph databases to the different fields of Molecular Biology exploring technical characteristics for the efficient storage of data. Finally, we conclude by discussing the role of NoSQL graph databases and the open field of research in this area.

2   NoSQL Graph Databases

Graphs naturally describe problem domains, and graph databases assemble simple abstractions of vertices and relationships in connected structures, making it possible to build models that are mapped closer to the problem domain. Often, data sets and their relationships are represented by graphs, and the importance of the information embedded in relationships has prompted an increase in the number of graph database initiatives [1]. This occurs due to various factors, such as the interests in recommending systems, circuits in engineering, social media, chemical and biological networks, and the search and identification of criminal cells [31].

Graph databases are Database Management Systems (DBMS) with Create, Read, Update, and Delete (CRUD) methods, which can store graphs natively or emulate them in a different database model [27]. The schema in graph databases can store data in vertices and, depending on the database, can also be stored in edges [30].

A significant aspect of graph databases is the way they manage relationships making it possible to establish them between entities. It is similar to storing pointers between two objects in memory. In addition, indexes can make the data retrieve of queries more efficient. However, there are some restrictions for types, as the BLOB type (Binary Large Object), which is not yet supported by graph databases.

3   Graph Databases Applied to Omics Data

In this section, we present works in which the NoSQL graph databases bring contributions to the Molecular Biology using omics data.

With the advent of NoSQL databases, a fundamental question loomed: would the NoSQL databases be ready for Bioinformatics? Have and Jensen [16] published a paper answering this question for NoSQL graph databases. In their work, they measured the performance of the graph database Neo4J v1.8 and the relational database PostgreSQL v9.05 executing some operations on data from STRING [36]. They found, for example, that the graph database found the best scoring path between two proteins faster by a factor of almost 1000 times. Also, the graph database found the shortest path 2441 times faster than the relational database when constraining the maximal path length to two edges. The conclusion was that graph databases, in general, are ready for Bioinformatics and they could offer great speedups on selected problems over relational databases.

Bio4j [26] proposes a graph-based solution for data integration with high-performance data access and a cost-effective cloud deployment model. It uses Neo4J to integrate open data coming from different data sources considering the intrinsic and extrinsic semantic features. Corbacho *et al.* [9] used the Bio4J graph database for Gene Ontology (GO) analyzes in *Cucumis melo*.

ncRNA-DB [6] is a database that integrates ncRNAs data interactions from a large number of well-established online repositories built on top of the OrientDB. It is accessible through a web-based platform, command-line, and the ncINetView, a plugin for Cytoscape[1], which is a software for analyses and visualization of biological networks. Another Cytoscape plugin is the cyNeo4j [33],

---

[1] www.cytoscape.org.

designed to link Cytoscape and Neo4j and enable an interactive execution of an algorithm by sending requests to the server.

Henkel *et al.* [17] used the Neo4J to integrate the data from distinct system biology model repositories. This database offers curated and reusable models to the community, which describe biological systems through Cypher Query Language - the native query language of Neo4J.

Lysenko *et al.* [20] used a graph database to provide a solution to represent disease networks and to extract and analyze exploratory data to support the generation of hypotheses in disease mechanisms.

EpiGeNet [4] uses the Neo4J to storage genetic and epigenetic events observed at different stages of colorectal cancer. The graph database enhanced the exploration of different queries related to colorectal tumor progression when compared to the primary source StatEpigen^2.

The Network Library [34] used Neo4J to integrate data from several biological databases through a clean and well-defined pipeline.

2Path [30] is a metabolic network implemented in the Neo4J to manage terpenes biosynthesis data. It used open data from several sources and was modeled to integrate important biological characteristics, such as the cellular compartmentalization of the reactions.

Biochem4j [35] is another work that seeks integration of open data from different sources using Neo4J. It goes beyond a database and provides a framework starting point for this integration and exploration of an ever-wider range of biological data sources.

GeNNet [11] is an integrated transcriptome analysis platform that uses Neo4J graph database to unify scientific workflows storing the results of transcriptome analyses.

BioKrahn [24] is a graph-based deductive and integrated database containing resources related to genes, proteins, miRNAs, and metabolic pathways that take advantage of the power of knowledge graphs and machine reasoning, to solve problems in the domain of biomedical science as interpreting the meaning of data from multiple sources or manipulated by various tools.

Messaoudi [23] evaluated the performance time needed for storing, deleting and querying biomedical data of two species: *Homo sapiens* as a large dataset and *Lactobacillus Rhamnosus* as a small dataset, using Neo4J and OrientDB graph databases. They found that Neo4J showed a better performance than OrientDB using 'PERIODIC COMMIT' technique for importing, inserting and deleting. On the other hand, OrientDB achieved best performances for queries when more in-depth levels of graph traversal were required.

Reactome [14] is a well-established open-source, open-data, curated and peer-reviewed database of pathways, which recently adopted the graph database as a storage strategy due to performance issues associated with queries traversing highly interconnected data. In this case, the adoption of graph database improved the queries reducing the average query time by 93%.

---

^{2} http://statepigen.sci-sym.dcu.ie.

Arena-Idb is a platform for the retrieval of comprehensive and non-redundant annotated ncRNAs interactions [7]. It uses two different DBMS: a relational MySQL and the graph database Neo4J, which is applied to handle the construction and visualization of the networks on a web page.

Table 1 summarizes the contributions of each reported work in this review. Although there are many NoSQL graph databases available, so far only three of them (Neo4J, OrientDB and Grakn) have been reported in this field as shown in the Fig. 1.

Table 1. Contributions of graph-oriented databases for Molecular Biology

Graph database	Main contribution	Other contributions	Source
Neo4J	Biological networks	Protein-protein interaction	[16]
Neo4J	Gene annotation	GO analyses	[9]
OrientDB	Data integration	ncRNA interactions	[6]
Neo4J	Data integration	-	[17]
Neo4J	Data integration	-	[26]
Neo4J	Data visualization	-	[33]
Neo4J	Biological networks	Diseases association	[20]
Neo4J	Cancer	Epigenetic events	[4]
Neo4J	Data integration	-	[34]
Neo4J	Biological networks	Metabolic networks	[30]
Neo4J	Data integration	-	[35]
Neo4J	Transcriptome analyses	-	[11]
Grakn	Data integration	Biomedical analyses	[24]
Neo4J/OrientDB	Biomedical analyses	-	[23]
Neo4J	Biological networks	Metabolic networks	[14]
Neo4J	Biological networks	ncRNA interactions	[7]

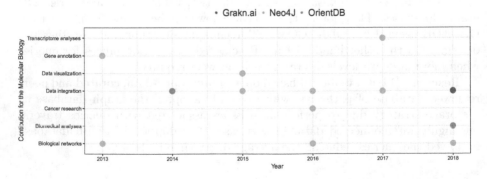

Fig. 1. Main contributions for the Molecular Biology using graph databases and omics data.

4   Discussion and Conclusion

In this work, we listed meaningful contributions of NoSQL graph databases to Molecular Biology using omics data. Performing queries across databases is routine activity in *in silico* biological studies, which, despite the available interfaces, is not a trivial task [35]. In this sense, the data integration is both a contribution to the field of Molecular Biology and Computer Science.

Data integration and biological networks were the most significant fields where the graph databases were employed. Data integration intends to represent relationships from previously related data knowledge, while metabolic networks intend to infer knowledge from existing relationships. However, it seems there is a hierarchy within data integration where it is a root contribution from which the others are derived. Biological networks are intuitively represented as graphs, and the use of graph databases for this purpose was predictable.

Once the data has already been processed and entered into the graph database, the queries become very intuitive and fast because of the way the nodes can be traversed. The performance and intuitiveness of queries in graph databases seem to be the main reason for using them as discussed in [14]. Graph queries are more concise and intuitive compared to equivalent relational database SQL queries complicated by joins. In addition, the engine of the graph databases is different, which leads to another point of investigation regarding the relationship between an engine and performance.

Databases contribute to the efficient storage of data, helping to ensure essential aspects of information security such as availability and integrity. The lack of schema in NoSQL graph databases, despite offering flexibility, can also remove the interoperability pattern of the data [20]. Graph database schemas may positively influence the maintainability of the graph databases, and open an ample field to examine the best graph schema for the data and their relationships concerning the normalization of data. A significant point to explore here is the threshold where the granularity of the vertices negatively influences the complexity and performance of the queries. Graph Description Diagram for graph databases (GRAPHED) [37] offers rich modeling diagrams for this purpose.

Although the scientific production using NoSQL databases is growing fast, the non-mutual citation supposedly shows a not explicit collaborative network. In summary, the use of NoSQL graph databases to store general data has increased, and the main contributions are related to data integration and performance in searches with queries traversing complex relationships. graph databases can help reach these solutions following the FAIR Guiding Principles for scientific data management and stewardship, which aims to improve the findability, accessibility, interoperability, and reuse of digital assets [40].

Acknowledgements. W. M. C. S. kindly thanks CAPES and IFG. M. E. M. T. W. thanks CNPq (Project 308524/2015-2).

References

1. Angles, R., et al.: Benchmarking database systems for social network applications. In: First International Workshop on Graph Data Management Experiences and Systems, p. 15. ACM (2013)
2. Bachman, C.W.: Integrated data store. DPMA Q. **1**(2), 10–30 (1965)
3. Bachman, C.W.: The origin of the integrated data store (IDS): the first direct-access dbms. IEEE Ann. History Comput. **31**, 42–54 (2009)
4. Balaur, I., et al.: EpigeNet: a graph database of interdependencies between genetic and epigenetic events in colorectal cancer. J. Comput. Biol. **24**, 969–980 (2017)
5. Berners-Lee, T., et al.: World-wide web: the information universe. Internet Res. **20**(4), 461–471 (2010)
6. Bonnici, V., et al.: Comprehensive reconstruction and visualization of non-coding regulatory networks in human. Front. Bioeng. Biotechnol. **2**, 69 (2014)
7. Bonnici, V., et al.: Arena-Idb: a platform to build human non-coding RNA interaction networks, pp. 1–13 (2018)
8. Codd, E.F.: A relational model of data for large shared data banks. Commun. ACM **13**(6), 377–387 (1970)
9. Corbacho, J., et al.: Transcriptomic events involved in melon mature-fruit abscission comprise the sequential induction of cell-wall degrading genes coupled to a stimulation of endo and exocytosis. PloS ONE **8**(3), e58363 (2013)
10. Corbellini, A., et al.: Persisting big-data: the NoSQL landscape. Inf. Syst. **63**, 1–23 (2017)
11. Costa, R.L., et al.: GeNNet: an integrated platform for unifying scientific workflows and graph databases for transcriptome data analysis. PeerJ **5**, e3509 (2017)
12. Crick, F.H., et al.: General nature of the genetic code for proteins. Nature **192**(4809), 1227–1232 (1961)
13. Deen, S.M.: Fundamentals of Data Base Systems. Springer, Heidelberg (1977). https://doi.org/10.1007/978-1-349-15843-0
14. Fabregat, A., et al.: Reactome graph database: efficient access to complex pathway data. PLoS Comput. Biol. **14**(1), 1–13 (2018)
15. Fry, J.P., Sibley, E.H.: Evolution of data-base management systems. ACM Comput. Surv. (CSUR) **8**(1), 7–42 (1976)
16. Have, C.T., Jensen, L.J.: Are graph databases ready for bioinformatics? Bioinformatics **29**(24), 3107 (2013)
17. Henkel, R., Wolkenhauer, O., Waltemath, D.: Combining computational models, semantic annotations and simulation experiments in a graph database. Database **2015** (2015)
18. Hutchison III, C.A.: Dna sequencing: bench to bedside and beyond. Nucl. Acids Res. **35**(18), 6227–6237 (2007)
19. Lander, E.S.: Initial sequencing and analysis of the human genome. Nature **409**(6822), 860–921 (2001)
20. Lysenko, A., et al.: Representing and querying disease networks using graph databases. BioData Min. **9**, 23 (2016)
21. Martin, R.G., et al.: Ribonucleotide composition of the genetic code. Biochem. Biophys. Res. Commun. **6**(6), 410–414 (1962)
22. McCallum, D., Smith, M.: Computer processing of dna sequence data. J. Mol. Biol. **116**, 29–30 (1977)
23. Messaoudi, C., Mhand, M.A., Fissoune, R.: A performance study of NoSQL stores for biomedical data NoSQL databases: an overview, November 2017 (2018)

24. Messina, A., Pribadi, H., Stichbury, J., Bucci, M., Klarman, S., Urso, A.: BioGrakn: a knowledge graph-based semantic database for biomedical sciences. In: Barolli, L., Terzo, O. (eds.) CISIS 2017. AISC, vol. 611, pp. 299–309. Springer, Cham (2018). https://doi.org/10.1007/978-3-319-61566-0_28

25. O'Neill, J.T.: MUMPS language standard, vol. 118. US Department of Commerce, National Bureau of Standards (1976)

26. Pareja-Tobes, P., et al.: Bio4j: a high-performance cloud-enabled graph-based data platform. bioRxiv (2015)

27. Robinson, I., Webber, J., Eifrem, E.: Graph Databases. O'Reilly Media Inc, Sebastopol (2013)

28. Sanger, F., Coulson, A.R.: A rapid method for determining sequences in DNA by primed synthesis with DNA polymerase. J. Mol. Biol. **94**(3), 441IN19447–441IN20448 (1975)

29. Shreeve, J.: The Genome War: How Craig Venter Tried to Capture the Code of Life and Save the World. Random House Digital Inc., Manhattan (2005)

30. Silva, W.M.C.D., et al.: A terpenoid metabolic network modelled as graph database. Int. J. Data Min. Bioinform. **18**(1), 74–90 (2017)

31. Srinivasa, S.: Data, storage and index models for graph databases. In: Sakr, S., Pardede, E. (eds.) Graph Data Management, pp. 47–70. IGI Global, Hershey (2011)

32. Stephens, Z.D., et al.: Big data: astronomical or genomical? PLoS Biol. **13**(7), e1002195 (2015)

33. Summer, G., et al.: cyNeo4j: connecting neo4j and cytoscape. Bioinformatics **31**(23), 3868–3869 (2015)

34. Summer, G., et al.: The network library: a framework to rapidly integrate network biology resources. Bioinformatics **32**(17), i473–i478 (2016)

35. Swainston, N., et al.: biochem4j: Integrated and extensible biochemical knowledge through graph databases. PloS ONE **12**(7), e0179130 (2017)

36. Szklarczyk, D., et al.: The string database in 2017: quality-controlled protein-protein association networks, made broadly accessible. Nucl. Acids Res. **45**(D1), D362–D368 (2017)

37. Van Erven, G., Silva, W., Carvalho, R., Holanda, M.: GRAPHED: a graph description diagram for graph databases. In: Rocha, Á., Adeli, H., Reis, L.P., Costanzo, S. (eds.) WorldCIST'18 2018. AISC, vol. 745, pp. 1141–1151. Springer, Cham (2018). https://doi.org/10.1007/978-3-319-77703-0_111

38. Venter, J.C., et al.: The sequence of the human genome. Science **291**(5507), 1304–1351 (2001)

39. Watson, J.D., Crick, F.H.: A structure for deoxyribose nucleic acid. Nature **171**(4356), 737–738 (1953)

40. Wilkinson, M.D., et al.: The FAIR guiding principles for scientific data management and stewardship. Sci. Data **3** (2016). https://doi.org/10.1038/sdata.2016.18

41. Wu, R., Taylor, E.: Nucleotide sequence analysis of DNA: II. Complete nucleotide sequence of the cohesive ends of bacteriophage λ DNA. J. Mol. Biol. **57**(3), 491–511 (1971)

ViMT - Development of a Web-Based Vivarium Management Tool

Cristiano Guimarães Pimenta[1], Alessandra Conceição Faria-Campos[1],
Jerônimo Nunes Rocha[1], Adriana Abalen Martins Dias[3],
Danielle da Glória de Souza[2], Carolina Andrade Rezende[2],
Giselle Marina Diniz Medeiros[2], and Sérgio Vale Aguiar Campos[1(✉)]

[1] Departamento de Ciência da Computação,
Universidade Federal de Minas Gerais, Belo Horizonte, Brazil
scampos@dcc.ufmg.br
[2] Biotério Central da Universidade Federal de Minas Gerais, Belo Horizonte, Brazil
[3] Departmento de Biologia Geral, Universidade Federal de Minas Gerais,
Belo Horizonte, Brazil

Abstract. Animal experimentation is still an important part in research
and experimentation, since it is still not possible to completely eliminate
animal testing. Therefore, it is necessary to find efficient ways to manage the processes that take place in animal facilities, thus aiding in the
achievement of the use of fewer animals and in more humane research
methods. Animals for research purposes are usually kept at vivariums.
One approach to help in the management of data in these facilities is the
use of Laboratory Information Management Systems (LIMS), specific
software for the management of laboratory information with emphasis on
quality. The present work describes ViMT, a LIMS designed to manage
the animal facility of the Federal University of Minas Gerais (UFMG),
Brazil. ViMT has been designed as a specialized LIMS having as major
objectives a flexible structure that makes it simple to model day by day
operations at a vivarium, trackability of operations and an easy to use
interface accessible from any computer or smartphone. ViMT has been
developed jointly with the UFMG Rodents Vivarium, and is currently
being deployed at this facility. It is a fully web-based, platform independent system and can be accessed using browsers.

Keywords: LIMS · Vivarium · Data management

1 Introduction

The advancement of research and development of new medical procedures
resulted in an increase in animal experimentation. According to the Royal Society for the Prevention of Cruelty to Animals, over a 100 million animals are
used in experiments every year around the world [14]. The most commonly used
species include rats, mice, guinea pigs, hamsters, rabbits, fishes, primates and
domestic animals [5]. These animals are kept at vivariums. A vivarium is the

© Springer Nature Switzerland AG 2018
R. Alves (Ed.): BSB 2018, LNBI 11228, pp. 58–65, 2018.
https://doi.org/10.1007/978-3-030-01722-4_6

place where live animals are kept for scientific research. It is built in a physical area of adequate size and divisions, where specialized personnel work. In this place there is water and food specific for each animal species, as well as constant temperature and appropriate artificial lighting.

In 1959, Russel and Burch [15] proposed a set of guidelines for more ethical use of animals known as the Three Rs(3Rs) that are also used at vivariums. The 3Rs are **R**eplacement: methods which avoid or replace the use of animals in research, **R**eduction: use of methods that enable researchers to obtain comparable levels of information from fewer animals, or to obtain more information from the same number of animals, and **R**efinement: use of methods that alleviate or minimize potential pain, suffering or distress, and enhance animal welfare for the animals used. The 3Rs have a broader scope than simply encouraging alternatives to animal testing. They aim to improve animal welfare and scientific quality where the use of animals cannot be avoided. Since the proposition of the 3 Rs several alternatives to animal experimentation have been developed, such as *in vitro* cell and tissue cultures, computer models and use of other organisms (e.g. microorganisms, invertebrates and lower vertebrates) [5,8]. Furthermore, adequate storage and exchange of information regarding animal experiments can greatly reduce the need for unnecessary repetitions, which would lead to fewer animals being used [3].

However, despite all the efforts to avoid it, animals are still used in toxicological screenings, studies related to the effects of medical procedures, drug development, testing of cosmetic products, environmental hazard identification and risk assessment [1,5,16]. Since it is still not possible to completely eliminate animal testing, it is necessary to find efficient ways to manage the processes that take place in animal facilities, thus aiding in the achievement of the 3Rs. One approach to achieve that is the use of Laboratory Information Management Systems (LIMS), which are computational systems used to track and manage laboratory information, generating consistent and reliable results [2,6,10].

The present work describes ViMT, a LIMS designed to manage the animal facility of the Federal University of Minas Gerais (UFMG), Brazil. ViMT has been designed as a specialized LIMS with the following objectives in mind: a flexible structure that makes it simple to model all day by day operations at a vivarium, trackability of operations and an easy to use interface accessible from any computer or smartphone.

ViMT has been developed jointly with the UFMG Rodents Vivarium, and is currently being deployed at this facility. When at full capacity, ViMT will manage all of the approximately 40,000 animals at the UFMG Vivarium.

2   Related Work

The management of the activities in a vivarium requires a very detailed and accurate record of all operations. The use of computational tools to aid in this task is well known. However, there are few specialized tools for this purpose, with some of them proprietary [11]. Mazzaroto and Silveira [9] have developed a free software tool for that – *BioterC* – to manage animals in a lab facility in Brazil. However, despite being designed specifically for vivariums, *BioterC* primarily manages stock control, animals orders and breeding, lacking the ability to track animal-related events, such as births, deaths and diseases, for example, which hinders trackability and makes the system inadequate for UFMG Vivarium.

The use of electronic notebooks and LIMS to manage laboratory data has been successfully reported but few of these tools provide the functions needed to manage vivarium's complex routines and most of them have been developed to fulfill the needs involved in their development and cannot be adapted to manage vivarium data [7,10]. For example, a biological laboratory typically receives a sample that is processed to generate other products, or to identify some property of the original sample. In a vivarium no such concept exists. The operations of a vivarium typically do not track a single entity (sample or animal), and are not related to its status in the system. Instead, they track sets of entities, cages of animals in our case, the operations being more similar to stock control systems than laboratory experiments.

ViMT has been designed to accommodate such flow of information, but at the same time it still allows individual animals to be tracked, maintaining trackability, a key requirement for vivariums and other laboratories. These features are often not present in systems not designed for this application.

3   Vivarium Operations

UFMG's Vivarium is responsible for the husbandry of several strains of mice and rats, following strict sanitary and ethical procedures, in order to guarantee a high genetic quality of the animals. The main activities of the laboratory are as follows: feed the animals, select animals for breeding, register births, wean the litter, track deaths and diseases, receive animal orders from researchers and deliver them.

In the UFMG Vivarium, animals are kept in individually ventilated cages, labeled with information about the rodent species, strain, number of animals in the cage, sex, date of birth, and other relevant dates such as breeding and weaning. The density of animals in the cage follows the recommended by the Guide of the Care and Use of Laboratory Animals [12]. A cage typically contains the parents and offspring of a single couple and is identified by a combination of the names of the parents' cages. A good record keeping and tracking system is crucial for the efficient management of the vivarium. For this, accurate cage/animal identification is necessary and all the procedures performed in a cage or in an individual animal must be trackable. It is also necessary to provide easy recovery of any information about each animal of the colony, including breeding, births, weaning and clinical findings such as illness, injury or death. The easy access to this set of data permits the monitoring of the productive rates, the detection of abnormalities and the improvement of the strategies of animal care, husbandry and production.

4   The ViMT System

Proper care, use, and humane treatment of animals used in research, testing, and education require scientific and professional judgment based on knowledge of the needs of the animals and the special requirements of the corresponding programs. Because of that ViMT has been developed through direct interaction with the vivarium personnel at UFMG. Therefore, ViMT is a tool that is adapted to the different routines in the vivarium and includes forms for information on the main activities performed there: the species and number of animals in the facility, the number of animals requested by researchers, housing and husbandry requirements, major operative procedures, removal, euthanasia and disposition of animals.

The system is divided into a server and a client applications. The server is implemented in Java and uses a MySQL relational database management system. The client consists of a web interface, which can be accessed via any modern browser, such as Google Chrome, Firefox, Safari and Opera. It was implemented using the Angular platform (https://angular.io) and a local database to allow offline access.

The workflow of the system is illustrated in Fig. 1. Each box represents an activity. UFMG Vivarium does not receive animals from external sources, so the only input is via births. This activity stores the number of animals that were born, the date, and the cage in which the birth occurred. The *breeding* activity stores the origin of the male and the female(s), the date and the identifier of the cage to which the animals were transferred. Breeding leads to new birth activities. The *separation* activity identifies the cage and the date. After breeding, the male is removed from the cage and euthanized. The offspring are weaned via the *weaning* activity and moved to a new box. It contains information regarding the origin and destination boxes, the number of animals of each sex that were weaned and the date. The output of the system is represented by the delivery, death or euthanasia of the animals. *Death* and *euthanasia* activities store the cage in which the event occurred, the number of animals of each sex, the cause and the date. *Delivery* contains the number of animals delivered, the cage from which they were removed and the date, and is related to *order*, which keeps track of the requesting researcher, their organization, the number of the ethics committee's protocol, the strain, sex, number and age of the animals and the payment method.

The system is available online at http://vimt.luar.dcc.ufmg.br. The tool is fully available online without the need for investment in new hardware, which simplifies data collection, since the software can be accessed from any equipment with Internet access, including smartphones. Furthermore, the use of ViMT provides a tool to simplify and improve data collection besides gains in the ability to manage inventories and increase the predictability in the provision and production.

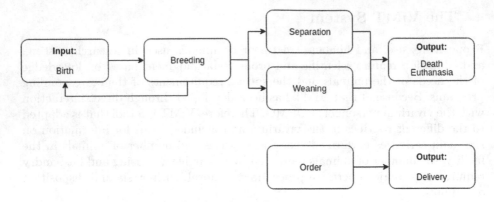

Fig. 1. ViMT's workflow.

4.1   Back-End

The back-end of the system uses the web server Apache Tomcat®.

Data Modeling. The database can be grouped into three functional categories: users, activities and attributes. Users can be one of two types: internal or external. Internal users correspond to the technicians (collaborators) and administrators of the animal facility, while external users are its clients. There is a permission system that grants users access to specific parts of the application depending on their type.

Activities are the central entity of the system, since they model all the animal-related functionalities. Each activity has a type, which defines the event associated to it (e.g. death, birth, breeding), and can be related to a parent activity.

They also have a set of attributes that define the activity. The structure and type of the attributes vary among the different types of activities. Examples of attributes include birth date, strain, order number. Attributes are modelled separately from activities, and are related to the activity that uses them. By modelling the system this way it becomes more flexible and easier to adapt to future modification in the system.

Client Integration. There are two communication interfaces between server and clients: a RESTful API and the WebSocket protocol. The RESTful API was implemented as Java Servlets and use CRUD operations, provided by the HTTP protocol [4], to allow clients to send requests to the server.

On the other hand, the WebSocket protocol provides full-duplex communication between the server and a client over a single TCP socket, which allows both ends to start the communication and decreases latency compared to HTTP [13]. This is important to allow multiple users to access the application concurrently. When the server detects an update, it sends a message to all connected clients with the new data, eliminating the need for HTTP polling.

Fig. 2. Collaborator dashboard.

Fig. 3. Animals page.

4.2 Front-End

The front-end comprises a client and a collaborator areas, which can only be accessed by users of the corresponding type. Clients are allowed to register themselves, but only administrators can create collaborator accounts.

Collaborator Area. The collaborator area is more complex than the client's, since they execute most of the activities. It has a *dashboard* that contains links to its functionalities: the panels for *Animals*, *Lineages* and *Orders* (Fig. 2). The *Animals* panel contains a list of the cages and information about the animals in each of them (Fig. 3). This page will also have options to search for specific cages besides filtering and sorting based on specific criteria.

The *Lineages* panel shows a list of all the animal strains maintained in the facility. Each item in the list is a link to the events page of the corresponding strain (Fig. 4), that allows the user to check the events registered for the strain

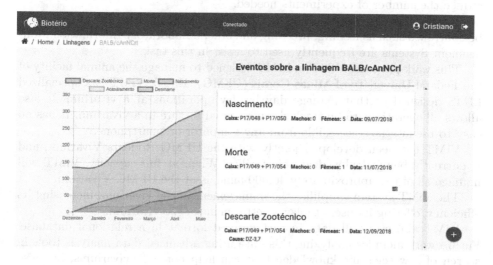

Fig. 4. ViMT Lineages panel showing the events registered.

Fig. 5. ViMT Orders panel showing the option to place a new order.

Fig. 6. ViMT Client area showing the option to place a new order.

and to add new events. There is also a graph summarizing the number of occurrences of each type of event over the last 6 months. The *Orders* panel exhibits the orders placed by the clients and the corresponding deliveries (Fig. 5).

Client Area. The client area (Fig. 6) is used for the external user (client) to follow the orders placed by him/her and the corresponding deliveries. It is also used by the client to place new orders.

5   Conclusion

Animal experimentation is an undesired but still necessary procedure. Millions of animals are used every year in experiments. These animals are kept at vivariums, where The quality, handling and use of these animals is controlled to increase the quality of the experiments in which these animals are used, and consequently reduce the number of experiments needed.

Maintaining a vivarium is complex, all aspects of the life of the animal must be controlled such as feeding, breeding and others. Laboratory Information Management Systems are frequently used to assist in this task.

This work proposes ViMT, a LIMS designed to manage the animal facility of the Federal University of Minas Gerais (UFMG), Brazil. ViMT is a specialized LIMS designed to that manage day by day operations at a vivarium. It also allows all operations to be tracked, an essential feature in a vivarium. It has an easy to use interface accessible from any computer or smartphone.

ViMT has been developed jointly with the UFMG Rodents Vivarium, and is currently being deployed at this facility. When at full capacity, ViMT will manage all of the approximately 40,000 animals at the UFMG Vivarium.

The ViMT system simplifies the management of the vivarium, increasing its efficiency, offering its users tools to better manage the vivarium.

ViMT also store all data in a structured format in a relational database. Future work includes analyzing this data using advanced data analysis tools in search of new scientific knowledge that can help not only vivariums, but also laboratories that use these animals.

6   Availability

The ViMT system can be accessed at http://vimt.luar.dcc.ufmg.br using the user "guest@vimt.com" and password "gu3st".

Acknowledgments. The authors wish to thank CNPQ, FAPEMIG, CAPES and Biotério Central da UFMG for financial support.

References

1. Aeby, P., et al.: Identifying and characterizing chemical skin sensitizers without animal testing: Colipa's research and method development program. Toxicol. Vitr. **24**(6), 1465–1473 (2010)
2. Avery, G., McGee, C., Falk, S.: Product review: implementing LIMS: a how-to guide. Anal. Chem. **72**(1), 57A–62A (2000)
3. Balls, M.: Replacement of animal procedures: alternatives in research, education and testing. Lab. Anim. **28**(3), 193–211 (1994)
4. Battle, R., Benson, E.: Bridging the semantic web and web 2.0 with representational state transfer (REST). Web Semant. **6**(1), 61–69 (2008)
5. Doke, S.K., Dhawale, S.C.: Alternatives to animal testing: a review. Saudi Pharm. J. **23**(3), 223–229 (2015)
6. Faria-Campos, A.C., et al.: FluxCTTX: a LIMS-based tool for management and analysis of cytotoxicity assays data. BMC Bioinform. **16**(Suppl 19), S8 (2015)
7. Hanke, L.A.: FluxTransgenics: a flexible lims-based tool for management of plant transformation experimental data. Plant Methods **10**(1), 20 (2014)
8. Liebsch, M., et al.: Alternatives to animal testing: current status and future perspectives. Arch. Toxicol. **85**(8), 841–858 (2011)
9. Mazzarotto, G.A.C.A., Silveira, G.F.: Desenvolvimento e implementação de um software livre para o gerenciamento de um biotério brasileiro **1**(2), 61–68 (2013)
10. Melo, A., Faria-Campos, A., DeLaat, D.M., Keller, R., Abreu, V., Campos, S.: SIGLa: an adaptable LIMS for multiple laboratories. BMC Genomics **11**(Suppl 5), S8 (2010)
11. Muench, M.O.: A cost-effective software solution for vivarium management. Lab. Anim. **46**(1), 17–20 (2016)
12. National Reasearch Council: Guide for the Care and Use of Laboratory Animals, 8th edn. The National Academies Press, Washington, DC (2010)
13. Pimentel, V., Nickerson, B.G.: Communicating and displaying real-time data with WebSocket. IEEE Internet Comput. **16**(4), 45–53 (2012)
14. RSPCA: The use of animals in research and testing (2018). https://www.rspca.org.uk/ImageLocator/LocateAsset?asset=document&assetId=1232742266055&mode=prd
15. Russell, W.M.S., Burch, R.L., Hume, C.W.: The Principles of Humane Experimental Technique (1959)
16. Scholz, S., et al.: A European perspective on alternatives to animal testing for environmental hazard identification and risk assessment. Regul. Toxicol. Pharmacol. **67**(3), 506–530 (2013)

An Argumentation Theory-Based Multiagent Model to Annotate Proteins

Daniel S. Souza[1]([✉])(ID), Waldeyr M. C. Silva[1,2](ID), Célia G. Ralha[1](ID), and Maria Emília M. T. Walter[1](ID)

[1] University of Brasília, Brasília, Brazil
dssouzadan@gmail.com, {ghedini,mariaemilia}@unb.br
[2] Federal Institute of Goiás, Formosa, Goiás, Brazil
waldeyr.mendes@ifg.edu.br

Abstract. Many computational and experimental methods have been proposed for predicting functions performed by proteins. *In silico* methods are time and resource-consuming, due to the large amount of data used for annotation. Moreover, computational predictions for protein functions are usually incomplete and biased. Although some tools combine different annotation strategies to predict functions, biologists (human experts) have to use their knowledge to analyze and improve these predictions. This complex scenario presents suitable features for a multiagent approach, e.g., expert knowledge, distributed resources, and an environment that includes different computational methods. Also, argumentation theory can increase the expressiveness of biological knowledge of proteins, considering inconsistencies and incompleteness of information. The main goal of this work is to present an argumentation theory-based multiagent model to annotate proteins, called ArgMAS-AP. Additionally, we discuss a theoretical example with real data to evaluate the suitability of our model.

Keywords: Protein function · Annotation · Multiagent systems · Argumentation theory

1 Introduction

Annotation of protein function consists in finding descriptions about how protein acts in its environment, under different conditions, e.g., heat, energy, interactions with/without water molecules, protein-protein interactions and post-translational modifications.

The broad concept of *function* does not have a widely accepted definition. Bork et al. [7] states that function should be described in different and specific contexts, e.g., molecular functions, cellular functions and phenotypic functions, including the interaction with the current environment conditions. Shrager [26] says that function may be described at different levels, from biochemical functions, biological processes and metabolic pathways, to organs and systems. Thus,

© Springer Nature Switzerland AG 2018
R. Alves (Ed.): BSB 2018, LNBI 11228, pp. 66–77, 2018.
https://doi.org/10.1007/978-3-030-01722-4_7

different ontologies and controlled vocabularies were defined to describe functions considering different contexts, e.g., Enzyme Commission (EC) [29], Gene Ontology (GO) [3] and Human Phenotype Ontology (HPO) [16].

The problem of protein function annotation consists in correctly assigning these descriptions to a not yet characterized protein. Different strategies were defined, from sequence level (i.e., DNA, mRNA and proteins in polypeptide chains) to structural (spatial) level. These strategies explore different biological concepts about proteins, e.g., homology, phylogeny, domains, active sites and molecular dynamics [6,22,28]. Analysis of these concepts gives biologists grounded knowledge about protein function, supporting their annotation task.

In general, *in silico* tools combine different strategies to improve their function prediction. Each strategy gives evidences about how protein acts. Evidences are used as possible hypothesis for the prediction, and have different levels of reliability, since they can either support the same prediction inferred by different tools or conflict with each other.

The annotation task, since requiring a deep knowledge about protein features and their biological roles, offers a complex scenario that is suitable for a multiagent environment [30]. Agents are designed to simulate different biologists' knowledge. These agents interact with each other, collaboratively and competitively, solving conflicts and reaching agreement about annotation evidence, so that the most plausible function for the analyzed protein can be predicted. On the other side, since evidences are hypothetical, many times supported with inconsistent and incomplete information, the annotation process could be improved with an argumentation model for solving conflicts. In particular, a multiagent system based on the argumentation theory [18] can be modeled in deliberation dialogues [17] that allow agents to construct an annotation agreement, by discussing and expressing their personal annotation preferences and opinions. In our model, deliberation dialogues based on expert knowledge formalized in the agents allow to reach a consensus, reinforcing the annotation task.

The objective of this work is to present an argumentation theory-based multiagent model for annotating proteins, called ArgMAS-AP. We also show a theoretical example designed to evaluate the suitability of the proposed model.

In Sect. 2, we briefly present biological and computational concepts, used in this work. In Sect. 3, we propose the argumentation theory-based multiagent model. In Sect. 4, we discuss the theoretical example with real data. Finally, in Sect. 5, we conclude and suggest future work.

2   Background

In this section, we briefly describe protein annotation strategies, multiagent system and argumentation theory.

2.1   Protein Annotation Strategies

Protein annotation *in silico* methods have explored biological data ranging from the sequence level of nucleotides to the structural/spatial level of proteins, including data from phylogeny, gene expression, protein-protein interactions and biomedical literature, each aspect unveiling a particular feature that contributes to find a protein function [21, 25]. Moreover, machine learning methods have been designed to integrate these features, aiming at improving function prediction performance. This provides stronger biological background, since it overcomes each method limitations, taking into account distinct strategies of annotation.

Differently from these methods, our multiagent system increases the expressiveness of protein biological knowledge, produced by different computational methods, treating inconsistencies and information incompleteness using the argumentation theory. An annotation for a protein is constructed based on a deliberation dialogue among agents, instead of providing inferences from either statistical measures or deductive reasoning. Besides, our model is flexible enough to integrate multiple annotation strategies, formalized with specialized agents. In this work, we are focused in sequence-based strategies, both BLAST [2], to compute similarity, and HMMER [11], to get information about domain architectures.

2.2   Multiagent System and Argumentation Theory

According to Weiss [30], intelligent agents are entities who pursue their goals, and perform tasks, such that their performance measures are optimized. These entities are flexible agents that behave rationally according to their environment conditions, limited by the available and acquired information and by their perception and action capabilities. A rational agent always attempts to optimize its performance measure. According to Wooldridge and Jennings [31], a flexible and autonomous behavior allows the agent to control features about its internal state, characterized as: (i) reactivity: agents presenting a stimulus-response behavior; (ii) proactivity: agents recognize opportunities and take initiatives; (iii) interactivity: through communication languages, agents show social skills, interacting with each other.

There are different reasoning models known in the literature to design intelligent agents [27]. One of them is Belief-Desire-Intention (BDI), with utility and goal driven-based agents. The BDI model is based on the theory of human practical reasoning, developed by Bratman [8], where the main focus is the role of intention in reasoning. Practical reasoning is directed by actions, or it is the process of deciding what to do based on three mental attitudes - beliefs, desires and intentions, respectively representing the informative, motivational and deliberative components of the agents [24].

A multiagent system (MAS) [30] is a collection of intelligent agents, each acting to reach his own (or common) goals, and they can interact in a shared environment with communication capabilities. In a MAS, coordination mechanisms are proposed to avoid states considered undesired for one or more agents, attempting to coordinate agents' goals and tasks. Two coordination mechanisms are: (i) cooperation: agents work collaboratively to increase their possibilities to reach shared goals; and (ii) competition: agents act against each other to maximize their own goals. Additionally, agents may have hybrid mechanisms, i.e., they may compete with each other, acting individually in the pursue of their personal goals, to collaboratively maximize the reach of their shared goals.

In a MAS, argumentation-based techniques can be applied to specify agents' autonomous reasoning at two levels [18]: (i) internally, e.g., beliefs' revision, decision making under uncertainty, and non-standardized preference policies; and (ii) externally, at communication level, with structured argumentation protocols that enable agents to expose their opinions and solve conflicts.

Argumentation is a verbal and social activity of reasoning, with the objective of increasing (or decreasing) the acceptability of a controversial point of view, for the listener or reader, using propositions that justify (or refute) this point of view through a rational judge [12]. Argumentation may be generally seen as a reasoning process, with four steps [9]: (i) building *arguments* (to support/against a "sentence") from the available information of the environment; (ii) determining *conflicts* among agents' arguments; (iii) evaluating *acceptability* of different arguments; and (iv) *taking conclusions* using arguments that justify them.

3   The ArgMAS-AP Model

In this section, we present ArgMAS-AP, an argumentation theory-based multi-agent model for annotating proteins (see Fig. 1). The annotation model explores two strategies (see Sect. 3.1), based on different aspects of the biological knowledge about proteins: (i) similarity of protein sequences; and (ii) conserved domain architecture similarity, of predicted protein domains (subsequences). The argumentation module, as described in Sect. 3.2, is responsible for the inference mechanism to suggest annotation, which uses deliberation dialogues among agents, based on the argumentation theory, to reach an annotation agreement, from the agents' annotation suggestions and explanations.

3.1   Annotation Strategy Module

In this module, the first step is to use each input (a protein sequence) as a query to BLAST [2] against the Swiss-Prot [4] database, to get information about at most the k-nearest neighbor similar sequences (statistically ranked), which may share common annotations with the input. Each of the similar sequences may be totally, incompletely or inconsistently[1] annotated. Annotations of protein names are retrieved from the UniProtKB Web Service[2], while annotations of the GO terms are retrieved from the UniProt-GOA [5] database.

Next, the similar proteins, together with the input, are queried to HMMER [11] against the Pfam-A [13] database, to calculate their corresponding domain architectures. This allows to measure the similarities of the architecture domains between each similar sequence with the input protein sequence.

Finally, all these predicted (and uncertain) information are analyzed in the argumentation module, which comprises:

---

[1] *Inconsistent* is a protein that has at least one incorrectly assigned annotation.
[2] https://www.uniprot.org/help/api.

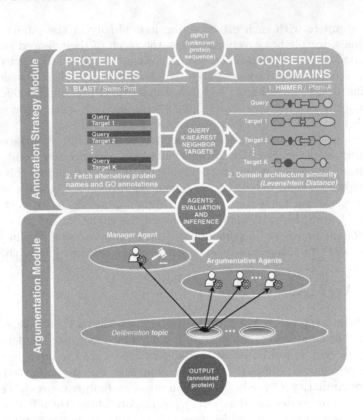

Fig. 1. High level scheme of the ArgMAS-AP model.

- a manager agent: that provides, for an input, one or more deliberation topics to be discussed, which are different aspects of annotation, e.g., protein name, molecular function, biological process or cellular component; and
- some argumentative agents: each agent generated from one similar protein (retrieved from BLAST and HMMER), with the objective of handling knowledge about its corresponding protein. Each agent participates in the discussion of a particular deliberation topic, by expressing/defending its protein's annotation suggestion, also arguing with the other agents, trying to reach a common agreement (the most plausible annotation).

At the end of the discussion of a topic, the manager evaluates which annotations are more plausible to be assigned to the input protein sequence. There are many possible deliberation scenarios. For example, two or more agents may either reach an annotation agreement by supporting each other, or conflict, by presenting distinct annotation suggestions. If one agent disagrees with the others, agents naturally engage in an argumentative dialogue, through a deliberation process, to find out which annotation suggestion leads to a mutual agreement. If any mutual agreement is possible among the agents, the manager agent evaluates the arguments, choosing one based on different levels of credulity [10], or accept all the conflicting arguments, since it cannot decide which one of the different suggestions is the best to annotate the input protein sequence.

Protein Sequence Strategy. There are no measures known in the literature that precisely relate sequence similarity to its functional similarity. The strategy proposed in this work is based on the Pearson's protocols of homology [22]. For *in silico* methods, sequence similarity is the most widely used and reliable strategy for protein annotation, usually using BLAST, where homology is inferred from similar protein sequences. By homology, proteins may either diverge (paralogs) or conserve (orthologs) their functions. Pearson (2015) discussed function divergence related to paralogy and orthology. He showed some cases that even diverged proteins may share high function conservation. Also, there are cases where proteins that present low sequence and structural similarity may still share similar functions, what happens when different non-homologous sequences converge.

Conserved Domain Strategy. Functional similarity was found among orthologs with different levels of divergence, by measuring the conservation of protein sequences' domain architectures. This indicates that function conservation between orthologs demands higher domain architecture conservation, when compared to some types of homologs [15]. Even so, the decreasing of functional similarity is weak, if compared to the increasing of sequence divergence among orthologs [1]. The domain architecture divergence are mainly caused by domain shuffling, duplication and deletion events.

Methods developed with this strategy explore the similarity among the domain architectures of the compared sequences. In this case, similarity is based on conservation of domain architectures, consisting in a measure that allows to infer that two protein sequences may conserve the same function by orthology. Considering biological aspects of evolution, proteins are functionally diverse due to mutations caused by selective pressure. Domain shuffling, deletion and insertion events are mutations that lead to diversification of proteins, as much as their functions. These events are naturally similar to the concepts applied by the Levenshtein edit distance [19], which includes edit operations (insertion, deletion and substitutions). This distance aims at finding the minimum number of edit operations required to transform one architecture into another. Based on this edit distance, we propose in Definition 1 our domain architecture similarity:

Definition 1. *Domain architecture similarity. Let* $A = \{a_1, \ldots, a_i, \ldots, a_n\}$ *and* $B = \{b_1, \ldots, b_i, \ldots, b_m\}$ *two ordered domain architectures, where* a_i *and* b_i *are protein domains, and each edit operation costs 1. The distance between two architectures* $LD(A, B)$ *ranges from 0 to* $max(|A|, |B|)$, *where 0 means that they are equal and* $max(|A|, |B|)$ *means that they are totally dissimilar. Based on* $LD(A, B)$, *the similarity function can be formulated as follows:*

$$Sim(A, B) = 1 - \frac{LD(A, B)}{max(|A|, |B|)}, \text{ if } max(|A|, |B|) > 0$$

where $0 \leq Sim(A, B) \leq 1$, $Sim(A, B) = 0$ means that both architectures are totally dissimilar, and 1 means that they are equal.

According to Finn et al. [14], the Pfam database should not have overlapping families and the existing ones were organized into clans. However, families from the same clan still share general functions, even if they have diverged into more specific ones. Although the degree of divergence between families from the same clan and among alignments could be used to weight the editing operations of our similarity measure (Definition 1), we are not penalizing them.

3.2 Argumentation Module

This module includes the model for deliberation dialogues, implemented in the Baidd tool by Kok [17]. The manager agent supervises and coordinates its argumentative agents, to evaluate and suggest the most plausible protein annotation. It instantiates k argumentative agents based on the protein sequence strategy. Each argumentative agent is able to suggest an annotation, to construct arguments to support, or refute, an existing argument.

A Model for Deliberation Dialogues. Consists of a dialogue among agents, who discuss a deliberation problem, through a topic language with options, goals and beliefs, as shown in Definition 2.

Definition 2. *A deliberation dialogue context consists of: (i) an ASPIC argumentation system [23]; (ii) a topic language L_t, with options $L_o \in L_t$, goals $L_g \in L_t$, and beliefs $L_b \in L_t$; (iii) a mutual deliberation goal $g_d \in L_g$.*

The manager agent is responsible to present deliberation topics for different annotation contexts. The argumentative agents have the mutual goal g_d of suggesting annotations for the input protein sequences. This dialogue is modeled in several steps, through a communication language, by the so-called *speech acts* [20], in the form of proposals, questions or arguments [17].

When a deliberation is presented, agents act in turns. In each turn, agents can perform one or more speech acts in the dialogue. Excluding the proposal step performed by the manager agent, each step has a target, which can be attacked, supported or surrendered by the agents, in an argumentative context. When no agent can formulate new arguments, the manager agent finishes the deliberation context, and evaluates the winning proposal (if there is one), based on the agent that provided the best utility measure (the most plausible protein annotation).

A Model for an Argumentative Agent. Based on Kok [17], each argumentative agent plays a role in the system, simulating an expert, in a particular deliberation topic. These roles describe the agents' obligations and desires, as part of their contexts in the dialogue. Each role consists of a set of options that an agent might know, and its own goals, which can be either mutual or selfish. Internally, argumentative agents are modeled according to the BDI model, with the support of the ASPIC inference engine, jointly implementing a reasoning based on epistemic logic for the formulation of practical arguments. Actions (in the construction of practical arguments) connects the mutual goal to the available proposed options, which are determined by the agent with higher utility. The utility measure is built based on the argument that fits better the agents' preferences, according to their beliefs, desires and intentions.

4   A Theoretical Example with Real Data

To illustrate the usefulness and richness of the argumentation module to annotate proteins, we created an example of a complex scenario, where argumentation takes place. This case includes a simulation of a dialogue among two argumentative agents, coordinated by their manager agent. Each argumentative agent has knowledge about one protein obtained from sequence similarity. Since both proteins have different annotations, each agent suggests a different biological role, and both agents will be engaged in an argumentative dialogue.

First, we describe data, parameters and the agents' simulation. After, we discuss the deliberation process during the simulation, presenting the agents' proposals and arguments, as well as the winning proposal.

4.1   Data, Parameters and Agents' Simulation

Following the ArgMAS-AP model, we gave as input the sequence P0ACC1 from *Escherichia coli*[3], annotated as "Release factor glutamine methyltransferase (prmC)". The parameters for the protein and conserved domain strategies were default, for both BLAST and HMMER. The obtained results, which generated two agents were:

- P0ACC1 (input protein sequence): presents an architecture of a single domain $\{PF13847\}$ from the clan CL0063;
- Q6F0I4[4]: annotated as "Release factor glutamine methyltransferase (prmC)", was retrieved from sequence similarity, presenting e-value = 1e−21, identity = 27%, coverage = 82%, an architecture of a single domain $\{PF05175\}$ from the clan CL0063, with $Sim(P0ACC1, Q6F0I4) = 1$; and
- Q92G13[5]: annotated as "Bifunctional methyltransferase (prmC/trmB)", was retrieved from sequence similarity, presenting e-value = 2e−44, identity = 35%, coverage = 91%, an architecture of two domains $\{PF05175, PF02390\}$ from the clan $CL0063$, with $Sim(P0ACC1, Q92G13) = 0.5$.

This information was passed to the argumentation module, where the manager agent used them to instantiate two argumentative agents: A for Q6F0I4 and B for Q92G13. The argumentative agents' settings are:

---

[3] http://www.uniprot.org/uniprot/P0ACC1.
[4] http://www.uniprot.org/uniprot/Q6F0I4.
[5] http://www.uniprot.org/uniprot/Q92G13.

- roles: the biological role of the designated protein, related with P0ACC1;
- options: the recommended name of the corresponding protein;
- goals: the mutual goal *"proteinName"* (no selfish goals were considered); and
- beliefs: each agent instantiates the same rule set (Listing 1), and adds distinct facts about the designated protein into its corresponding belief base.

```
1  /* Agent's personal facts related to its designated protein. */
2  identity, e-value, coverage, domain architecture similarity - Sim.
3  /* Rules driven to the mutual goal proteinName
4     Supporting arguments */
5  [r1] proteinName <- annotation(X), identity, coverage.
6  [r2] proteinName <- annotation(X), Sim(P0ACC1, X) >= 0.5.
7  /* Counter-arguments */
8  [r3] not proteinName <- your identity <= 30 is in the twilight zone.
9  [r4] not proteinName <- your Sim(P0ACC1, Y) is less than mine.
10 [r5] not proteinName <- my identity is less than yours,
11     but I present a high coverage considering my Sim(P0ACC1, X) >= 0.5,
12     supporting that I may have better conservation.
```

Listing 1: Argumentative agents' belief base.

Next, the manager agent imposes the deliberation topic *"annotation(T)"* to be discussed, and also the mutual goal *"proteinName"* to be reached by its argumentative agents. Following, both agents join the dialogue, arguing each other during the deliberation process. In the last step, the manager agent evaluates the winning proposal, as shown in Fig. 2.

4.2 Discussion

This scenario shows an example where traditional methods usually fail in transferring annotation. The best statistically similar proteins can be miss-annotated, since they cannot have the same functions. This degree of uncertainty was exploited by the agents of ArgMAS-AP, using deliberation dialogue based on argumentation theory to increase the expressiveness of biological knowledge retrieved by two sequence-based annotation strategies: (i) inference by homology based on sequence similarity; and (ii) inference by orthology based on domain architecture similarity.

According to Fig. 2, the similar proteins Q6F0I4 and Q92G13 that generated the argumentative agents A and B, respectively, have different biological roles, which enabled both agents to engage in an argumentative dialogue. Even if Q6F0I4 had showed less conservation and coverage than Q92G13, and its similarity is in the "twilight zone" (*identity* $\leq 30\%$) [22], Q6F0I4 still stands for a high value of coverage with 100% of domain architecture similarity, considering that both input (P0ACC1) and Q6F0I4 have different domain families belonging to the same clan, which still conserve a more general function. Moreover, both proteins give hypothetical functions to P0ACC1, but they are still uncertain.

The expressiveness of knowledge about Q6F0I4 gives to agent A plausible reasons that overcome agent B counter-arguments, ensuring that its proposal

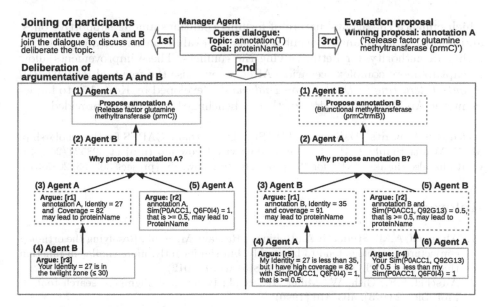

Fig. 2. Simulation of a deliberation dialogue of agents A and B, where green boxes represent the winning argument, and dashed red boxes represent a defeated argument. The winning proposal is the annotation of agent A, "Release factor glutamine methyl-transferase (prmC)".

may lead to a "valid" or plausible annotation. On the other side, the knowledge about Q92G13 is not strong enough to overcome argument A counter-arguments, leading to an inconsistent and defeated proposal. Therefore, the winning proposal taken by manager agent is the Q6F0I4 protein name, exactly the same name that was associated to the input sequence P0ACC1.

5 Conclusion

In this work, we proposed ArgMAS-AP, an argumentation theory-based multiagent model to annotate proteins. We also discussed a theoretical example, which demonstrated the suitability and feasibility of ArgMAS-AP. The deliberation dialogue between two agents improved knowledge, under uncertainty, about the protein given as input, mimicking the process normally done by biologists (human experts) to annotate proteins. Our simulation showed the richness of the argumentation theory, when applied to a complex scenario, where agents engage in a deliberation process, exposing and solving conflicts about their justified arguments, to reach a common agreement about the function of the input protein sequence.

Next steps are to refine the argumentation module as follows. The argumentative agents reasoning and knowledge can be improved: defining more complex rules, considering their utility measure; adding heuristics to belief revision, which enhance the agents' nature of forming coalitions among supportive ones; and adding to the manager agent a consistent evaluation strategy, e.g., voting, utility or authority, for better solving the conflicts. These improvements allow to capture more complex scenarios. After, we will use the free tool BAIDD *BDI Agents Interacting in Deliberation Dialogues*[6], developed by KoK [17], to implement the ArgMAS-AP model. Further, a benchmark can also be provided.

Acknowledgments. D. Souza and W. Silva kindly thank CAPES for the scholarship. M. E. Walter thanks CNPq for the productivity fellowship (project 308524/2015-2). C. Ralha also thanks CNPq for the productivity fellowship (project 303863/2015-3).

References

1. Altenhoff, A.M., Studer, R.A., Robinson-Rechavi, M., et al.: Resolving the ortholog conjecture: orthologs tend to be weakly, but significantly, more similar in function than paralogs. PLOS Comput. Biol. **8**(5), 1–10 (2012)
2. Altschul, S.F., Gish, W., Miller, W., et al.: Basic local alignment search tool. J. Mol. Biol. **215**(3), 403–410 (1990)
3. Ashburner, M., Ball, C.A., Blake, J.A., et al.: Gene ontology: tool for the unification of biology. Nat. Genet. **25**(1), 25–29 (2000)
4. Bairoch, A., Apweiler, R., Wu, C.H., et al.: The universal protein resource (uniprot). Nucl. Acids Res. **33**(Suppl. 1), D154–D159 (2005)
5. Barrell, D., Dimmer, E., Huntley, R.P., et al.: The goa database in 2009-an integrated gene ontology annotation resource. Nucl. Acids Res. **37**(Suppl. 1), D396–D403 (2009)
6. Bernardes, J.S., Pedreira, C.E.: A review of protein function prediction under machine learning perspective. Recent Pat. Biotechnol. **7**(2), 122–141 (2013)
7. Bork, P., Dandekar, T., Diaz-Lazcoz, Y., et al.: Predicting function: from genes to genomes and back. J. Mol. Biol. **283**(4), 707–725 (1998)
8. Bratman, M.E.: Intentions, Plans, and Practical Reasoning. Harvard University Press, Cambridge (1987)
9. Caminada, M., Amgoud, L.: On the evaluation of argumentation formalisms. Artif. Intell. **171**(5–6), 286–310 (2007)
10. Charwat, G., Dvořák, W., Gaggl, S.A., et al.: Methods for solving reasoning problems in abstract argumentation a survey. Artif. Intell. **220**, 28–63 (2015)
11. Eddy, S.R.: Profile hidden Markov models. Bioinformatics **14**(9), 755–763 (1998)
12. van Eemeren, F.H., Grootendorst, R., Henkemans, A.F.S., et al.: Fundamentals of Argumentation Theory. Handbook of Historical Backgrounds and Contemporary developments, 1st edn. Lawrence Erlbaum Associates, New Jersey (1996)
13. Finn, R.D., Bateman, A., Clements, J., et al.: Pfam: the protein families database. Nucl. Acids Res. **42**(Database issue), D222–D230 (2014)
14. Finn, R.D., Mistry, J., Schuster-Böckler, B.: Pfam: clans, web tools and services. Nucl. Acids Res. **34**(Suppl. 1), D247 (2006)
15. Forslund, K., Pekkari, I., Sonnhammer, E.L.: Domain architecture conservation in orthologs. BMC Bioinf. **12**(1), 326 (2011)
16. Khler, S., Vasilevsky, N.A., Engelstad, M., et al.: The human phenotype ontology in 2017. Nucl. Acids Res. **45**(D1), D865 (2017)

---

[6] https://bitbucket.org/erickok/baidd.

17. Kok, E.M.: Exploring the practical benefits of argumentation in multi-agent deliberation. Ph.D. thesis, Utrecht University (2013)
18. Maudet, N., Parsons, S., Rahwan, I.: Argumentation in multi-agent systems: context and recent developments. In: Maudet, N., Parsons, S., Rahwan, I. (eds.) ArgMAS 2006. LNCS (LNAI), vol. 4766, pp. 1–16. Springer, Heidelberg (2007). https://doi.org/10.1007/978-3-540-75526-5_1
19. Navarro, G.: A guided tour to approximate string matching. ACM Comput. Surv. **33**(1), 31–88 (2001)
20. Oishi, E.: Austins speech act theory and the speech situation. Esercizi Filosofici **1**(1), 1–14 (2006)
21. Pandey, G., Kumar, V., Steinbach, M.: Computational approaches for protein function prediction: a survey. Technical report, Department of Computer Science and Engineering, University of Minnesota, Twin Cities (2006)
22. Pearson, W.R.: Protein function prediction: problems and pitfalls. Curr. Protoc. Bioinf. **51**, 4–12 (2015)
23. Prakken, H.: An abstract framework for argumentation with structured arguments. Argum. Comput. **1**(2), 93–124 (2010)
24. Rao, A.S., George, M.P.: BDI agents: From theory to practice. In: First International Conference on Multi-Agent Systems (ICMAS-95) (1995)
25. Shehu, A., Barbará, D., Molloy, K.: A survey of computational methods for protein function prediction. In: Wong, K.C. (ed.) Big Data Analytics in Genomics, pp. 225–298. Springer, Cham (2016). https://doi.org/10.1007/978-3-319-41279-5_7
26. Shrager, J.: The fiction of function. Bioinformatics **19**(15), 1934–1936 (2003)
27. Stuart Russell, P.N.: Artificial intelligence: a modern approach. Prentice Hall Series in Artificial Intelligence, 3rd edn. Prentice Hall, Upper Saddle River (2010)
28. Tiwari, A.K., Srivastava, R.: A survey of computational intelligence techniques in protein function prediction. Int. J. Proteomics **2014**, 22 p. (2014). https://doi.org/10.1155/2014/845479
29. Webb, E.C.: Enzyme nomenclature. Recommendations of the Nomenclature Committee of the International Union of Biochemistry and Molecular Biology on the Nomenclature and Classification of Enzymes. Elsevier Inc., Academic Press, Cambridge (1992)
30. Weiss, G.: Multiagent Systems: A Modern Approach to Distributed Modern Approach to Artificial Intelligence, 1st edn. The MIT Press, Cambridge (1999)
31. Wooldridge, M., Jennings, N.R.: Intelligent agents: theory and practice. Knowl. Eng. Rev. **10**, 115–152 (1995)

AutoModel: A Client-Server Tool for Intuitive and Interactive Homology Modeling of Protein-Ligand Complexes

João Luiz de A. Filho[1], Annabell del Real Tamariz[2],
and Jorge H. Fernandez[1(✉)] (iD)

[1] LQFPP, Center of Biosciences and Biotechnology,
State University of North Fluminense, Campos dos Goytacazes, R.J, Brazil
joaoluiz.af@gmail.com, jorgehf@uenf.br
[2] LCMAT, Center of Tecnological Sciences,
State University of North Fluminense, Campos dos Goytacazes, R.J, Brazil
annabell.brasil@gmail.com

Abstract. The protein tertiary structure prediction is not a simple task but the assessment of this information becomes essential for functional annotation. Computer protein structure prediction is an important tool in structural biology helping to construct large quantity of interaction model of protein complexes or used to obtain three-dimensional structure and functional information of non-crystalizing proteins. However, the complexity of modeling softwares and a hard-to-use user interface makes it difficult the use for non-expert scientists. On this context, semi-automatic client-server software for protein homology modeling was developed, the AutoModel. The main goal of AutoModel is to provide a graphical, intuitive, interactive and practical interface to perform modeling experiments in a distributed architecture, with the possibility of importing water and ligand structural information from pdb templates, intended for easy modeling of different protein-ligand complexes. Our system facilitates the new users interaction with the modeling pipeline as it follows: 1. Searching structural templates; 2. Sequences Alignment; 3. Protein modelling; 4. Model refinement and 5. Loops refinement. In AutoModel 0.5 development we evaluated the use of different alignment tools in order to increase the quality of generated models, reduce the computational cost, and evaluate the impact of these changes in modeling quality and the experimentation speed. Our data suggest that using Muscle as alignment tool in the pipeline increases the quality of obtained models if compared to the other tested releases with significantly lower computational costs, which is always interesting in a distributed system running on a central server as AutoModel. "AutoModel Server" and "AutoModel Client" packages are available for Linux users through pypi package index. AutoModel is also freely available for academic community "as is" in http://biocomp.uenf.br.

Keywords: Protein structure prediction · Homology modelling
Client-server architecture · Modeller · AutoModel

Electronic supplementary material The online version of this chapter (https://doi.org/10.1007/978-3-030-01722-4_8) contains supplementary material, which is available to authorized users.

R. Alves (Ed.): BSB 2018, LNBI 11228, pp. 78–89, 2018.
https://doi.org/10.1007/978-3-030-01722-4_8

1 Introduction

Protein folding is an important biological question, concerning biophysics, biology and evolution and reflecting in the central goal of their proper functional characterization. Although the proteins structure may be divided into four hierarchical organization levels, where the higher level depends on the lower level of information [1], the assessment of the protein three-dimensional (3D) structure becomes essential for functional annotation and application of this knowledge on rational drug design [2–4].

In this context, predicting the protein tertiary structure is not a simple task and traditionally experimental methods used for this purpose (nuclear magnetic resonance (NMR), and crystallography mixed with X-ray diffraction) produce accurate models with atomic resolution [5]. However, the NMR has difficulty in determining the proteins structure which has more than 120 amino acid residues and some proteins such as membrane proteins do not crystallize satisfactorily, not allowing the use of X-ray diffraction [6]. Moreover, both methods are costly in time and resources and therefore difficult to apply on a large scale [5, 7] and resulting in an exponentially growing gap between sequenced proteins and their structures.

Thus, computational methods are being developed for determining the 3D structure of a protein using only its amino acid sequence and implementing a computationally efficient algorithm for the simulation of all the physical forces involved in folding proteins on water [7]. Even currently one of the biggest challenges on structural biology now, at this moment only small peptides or domains were determined properly using this "*ab initio*" approach [8, 9]. Analyzing the CASP 12 results [10], current development seems to consider the mixing of "*ab initio*" methods with other methodologies such as threading and sequence-based homology [8, 11], but this methodology that develops so fast will not be the center of considerations in this document. However, other methodological approach, the comparative or homology modeling may provide a useful 3D model for a protein [12, 13].

1.1 The Homology Modeling Prediction Method

The homology modeling prediction method relies on the assumption of structural similarity based on sequence similarity, and attempts to solve the protein 3D structure (target protein) using structural information from one or many proteins that works as structural templates. These templates are proteins that must have two characteristics: (*i*) known structure solved by experimental methods and (*ii*) be "homologous" to target protein. As "structural or sequential homology" is difficult to define, the identity of protein target and template sequences is a big determinant of this modeling methodology, being resolved in the sequence alignment step. These features are used by the modeling software for the creation of three-dimensional model. The homology modeling usually works as a pipeline and is divided into four distinct stages: (1) finding and selection of templates protein, (2) sequence alignment, (3) building the 3D structure of the model based on the alignment and (4) errors prediction and model validation [14]. The general pipeline is represented in Fig. 1A.

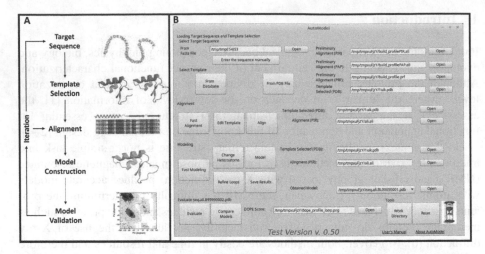

Fig. 1. The classical protein homology modeling pipeline as implemented in AutoModel. A: The algorithm requires a sequence of interest (target) and a template protein with structural information. After a series of steps, the three-dimensional structure of the target protein is obtained. B: Central window of AutoModel 0.5. The main window of AutoModel is divided in five fields corresponding to every stage of homology modeling pipeline.

This method can be performed using specific software for each step, for example, the HMMER [15] or Clustal [16] software to perform the alignment step and PRO-CHECK that can be used to validate the model stereo-chemical quality [17]. However, the increased demand for complete modeling tools turns the engine for the development of complete and professional program suits capable of template searching, multiple sequence alignment, structure modeling, molecular simulations, functional characterization, docking experiments and etc. Shrodinger [18] and Tripos [19] may illustrate these excellent but expensive to nonspecialists solutions. A different but common implementation of modeling programs as automatic web resource makes it easy to use the resource for many users. Example of such implementation are the SWISS-MODEL [20], ModWeb [14], Mholline [21] and ESyPred3D [22]. Although these services facilitate the daily use of basic homology modeling with excellent quality results, in general they perform an automatic internal pipeline and do not allow any user intervention in modeling parameters, do not import to the model structural water or even ligands complexed with structural templates. Furthermore, other programs provide a full interactive pipeline for homology modeling experiments but their only-text interface becomes a difficult barrier to overcome for a non-expert users. This is the case of Modeller program [17].

Modeller is a program for comparative protein structure modeling [23]. This program calculates the atomic coordinates of all non-hydrogen atoms tacking as an input to the target protein sequence, the structure of the template and the alignment of these sequences [13]. Modeller may also perform auxiliary tasks including structural profile building [24], structural template searching, multiple alignments, phylogenetic trees calculations and specific loop re-modeling in protein structures [23]. Although containing a lot of resources for homology modeling experimentation, Modeller undergoes on absence of intuitive graphical interface, making the learning process in a day-by-day experimentations a painful process for non-experts researchers.

In this context, we developed the AutoModel, an online tool that allows new or unexperienced user to make predictions of the three-dimensional proteins structure using the homology modeling method (Fig. 1B). The AutoModel differs from other tools for performing user-interactive and semi-automatic sessions, i.e., allowing the user to change important parameters in the prediction process when necessary. AutoModel code is freely available for academic community "as is" in http://biocomp.uenf.br.

2 Implementation

2.1 Inside the AutoModel Architecture

The AutoModel was developed following the classical Client-Server architecture. Briefly, the AutoModel Server waits connections and requisitions of AutoModel Client through internet. Any interested user must install the Client software to perform a modeling session. Every modeling session opens a request and AutoModel Client sends the necessary files and data to AutoModel Server for intensive processing. In turn, AutoModel Server processes this data using an appropriate python script, PDB database, Modeller program and other resources (Fig. 2). At the end of the desired calculation, the server sends the result to the client. From the user perspective, Auto-Model runs locally but all intensive tasks run on AutoModel Server. The AutoModel Server is based in a dual Xeon E56XX hardware (16 CPUs), can process up to 64 requests and allows multiple connections of AutoModel Clients simultaneously (Fig. 2).

2.2 The AutoModel Server

The AutoModel Server hardware is installed in a Linux server located at the Laboratório de Química e Função de Proteínas e Peptídeos, Centro de Biociências e Biotecnologia at the Universidade Estadual do Norte Fluminense (LQFPP-CBB-UENF). It was written in language python version 2 and uses reports of the obtained model is generated using Procheck. Modeller provides a python API that facilitates the development of all modules in python (Fig. 2).

Each module is a python script and must have a "create_script_in_folder()" and "run()" methods. When AutoModel Client makes one request, the AutoModel Server receives all the desired commands and users files necessary for operation. The Auto-Model Server then creates a temporary folder and one script that execute all the necessary third-party programs. After execution, the resulting files are sent back to the AutoModel Client. During the session, this operation is completely transparent for the user. Furthermore, the input-output formats from one step to the other inside the modeling pipeline require some format conversion (FASTA to PIR and etc.) and for this purpose the BioPython library [25] was used.

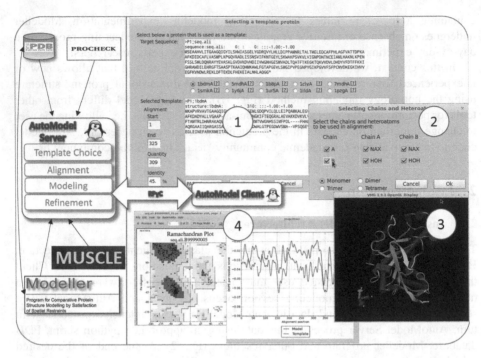

Fig. 2. General architecture of AutoModel. AutoModel Server contains the Modeller, MUSCLE and Procheck programs, PDB database and several other modules for each stage of the modeling pipeline. Several AutoModel Clients can connect to the server via internet using RPyC library and perform an interactive session of template search (1), sequence to structure alignment (2), modeling (3) and model refinement (4).

2.3 The AutoModel Client

The AutoModel Client provides a graphical interface designed to facilitate non-expert users to perform the homology modeling experiment in intuitive and simple graphical environment. This software was written in python language version 2 and was designed in two layers. The first layer connects to the server and performs all network requests operations in the AutoModel Server through RPyC library [26], controlling the modeling procedure. At this point all details of modeling experiment are managed by the user through the AutoModel Client and the AutoModel Server is only used when high CPU usage, third-part programs or database searching is necessary. The second layer in AutoModel Client was written using WxPython [27] library and is in charge of generating the graphical interface for the user and receive user commands (Fig. 2). Therefore, AutoModel Client was designed to use very low computer resources and avoid installation of third-part programs. Although the modular architecture and the use of python language and RPyC and WxPython libraries in theory allow the installation of AutoModel Client in Windows and OSx operating systems, this release of Auto-Model Client was tested only in Linux operating systems and is recommended for Ubuntu and Mint distributions.

3   Results and Discussion

Modeling session with Automodel is fully interactive and may be repeated in loop, the user conducts sessions in a semi-automatic way. This characteristic of the AutoModel allows the non-expert users to test pros and cons in every step of the experimental pipeline and perform modeling experiments with low or no previous knowledge on structural Computational Biology and this becomes the central goal of the described program herein.

3.1   AutoModel Interface and Normal Modeling Session

The implemented interface of AutoModel for the user is divided in five sections (Fig. 1 and 2), matching the 5 steps of homology modeling pipeline: (*i*) Template identification, (*ii*) sequence to structure alignment, (*iii*) protein modeling, (*iv*) model assessment, validation and refinement. In each step the user can change parameters that directly influence in quality of generated model or in the specific experimental needs. In order to use AutoModel, interested user only needs the sequence of the protein of interest. Typically, target sequence is provided to the program in FASTA format.

After loading the target sequence, the user must search a template protein that has a known three-dimensional structure. The AutoModel allows to load a known protein or to select it from a list of homologous sequence and this template may be used for the entire experimental pipeline. At this moment, it is only possible to select one template sequence (Fig. 2-1) but next generation of the program will provide the user with the possibility to build a structural profile of several structures and use this information in the modeling procedure. Subsequently, the user can select specific chains, structural waters and several heteroatoms or ligands from the template of choice. All these important components if available in the template may be used to the construction of the desired structural model. Furthermore, the heteroatoms or ligands selected in one experiments can be quickly replaced by others before the next step of the model building. The necessary alignment between sequences is performed on pressing only the "Align" button of the graphical interface (Fig. 2-2). In order to perform the alignment step, the AutoModel 0.5 uses the Muscle software, providing fast and accurate results. Resulting pairwise alignment will be used in the next step of the modeling pipeline: the structural model construction.

Three-dimensional model construction step is done by pressing the button "Get Model", this instructs the AutoModel to construct the protein model using the target sequence, the template of choice with heteroatoms and waters, and the alignment using the Modeller software (Fig. 2-3). In the AutoModel server five models will be constructed and the one with lower scoring energy function will be presented to the user as the best result of the modeling procedure. In the header of each pdb file generated by Modeller is the calculated "modeller objective function", the parameter used by AutoModel to select the better model.

To assess the general stereo-chemical quality of the generated model, AutoModel presents to the user two options: (*i*) calculate the Discrete Optimized Protein Energy (DOPE) per residue of target model and template structure and compare both in a simple graphic or (*ii*) general stereo-chemical analysis of obtained model in

PROCHECK program (Fig. 2-4). DOPE score is a statistical potential generated by Modeller software in three dimensional construction of the model [28]. To perform the second option, AutoModel Server uses the software Procheck, which generates several files with detailed stereo-chemical analysis and comparative quality of the obtained model. The data generated by the model assessment may be necessary in the refinement step. In AutoModel, the user may refine specific loops or detect regions of model with bad energy. To perform this refinement, the user may select the "Loop Refinement" option to select the loop region that will be refined. Made this, AutoModel uses specific routine of the Modeller program to refine the desired loop region of model. At the end of "loop refinement" procedure, a new model is generated, may be assessed using the previously described tools and refined in other parts if necessary.

The RPyC library allows the development of the softwares using object-proxying, a technique that permits to manipulate remote objects as if it was local. To do it, the RPyC library can be arranged to use synchronous and asynchronous operations. On the synchronous systems when the client software accesses remote objects, it is necessary to wait until the end of the operation and only before that, the client can continue the procedures. On the asynchronous operations, the client software can access remote objects or send a batch of works to the server and continue with his local operation. We tested the AutoModel using synchronous and asynchronous operations (Fig. 3). On the synchronous operations, all modeling data were in AutoModel Server. We realized that this design raised the volume of data necessary to complete the modeling process (Fig. 3). With this, the AutoModel was unstable on slow internet connection as mobile devices or the old infrastructure used in most developing countries. Indeed, implemented AutoModel using the asynchronous operation was more responsible, stable and faster. Moreover, it was possible to use a design where all user data of modeling pipeline was stored in AutoModel Client, thereby reducing the data volume necessary to complete the experiment, but increasing client hardware needs.

3.2   Comparison of Alignment Procedures and Modeling Study Case

To better exemplify the presented here AutoModel 0.5 release, a set of different proteins was used to access the quality of the generated models, the speed of the service and the impact of the alignment tool used in the general result of the homology modeling experiment (Table 1 and Figs. S1–S3).

For theses comparative experiments tree different proteins were used. First, the modeling of lactate dehydrogenase from *T. vaginalis* (TvLDH) [GenBank: AF070994] protein using as a template the 1bdm pdb [29] (45% of sequence identity; [13, 30]), as the control experiment used in the general example of the modeler webpage [31] (Table 1 and Fig. S1). As a second example, the homology modeling of the storage 7S vicilin from *A. angustifolia* [GenBank: AAM81249.1] with the 1uij:A pdb [32] as a template (29% identity; Table 1 and Fig. S2). The last modeled protein was the (putative) developmental protein cactus from *A. aegypti* [refseq:XP_001650267] with the 1iknD pdb as a template (40% sequence identity; Table 1 and Fig. S3). All these modeling experiments were performed in a low sequence identity for better understanding of the impact of the alignment tool in the modeling procedure.

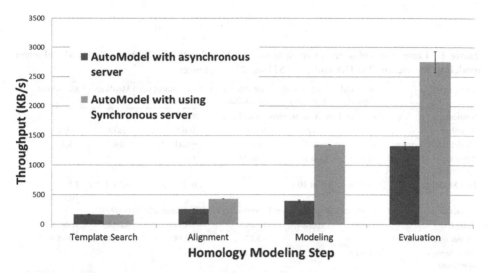

Fig. 3. Comparison between asynchronous and synchronous implementation of server-client communication in AutoModel.

In AutoModel 0.5 development, we evaluated the use of different alignment tools in order to increase the quality of generated models, reduce the computational cost, and evaluate the impact of these changes in modeling quality and speed of the experimentation. Modeller alignment algorithm of 9v4 and 9v9 releases, HMMER and Muscle programs were used with the same protein set. Thus, we evaluated: (i) the quality of the generated models using the Prosa-web Server; (ii) machine time for full modeling experiment and (iii) the time lapse of the alignment step. Our data suggest that the use of AutoModel with the 9v9 Modeller release obtained better results than generated models by Modeller 9v4 and the HMMER releases of the program pipeline, demonstrating a dramatic improvement in alignment algorithm of 9v9 release. For the overall quality of the generated model, the quality score of the external PROSA server was used [33] (Table 1 and Figs. S1–S3).

The general analysis of the obtained results is represented in Table 1 and Figs. S1 to S3, and points to a better performance when alignment step of the modeling pipeline is performed by the MUSCLE algorithm, showing a better balance in experimental speed and accurate results. At the same time, it is important to remark that the last release of Modeller (9v9) improved the speed of the alignment procedure and the quality of the obtained results, a problem detected in the 9v4 release of the program. However, using Muscle as alignment tool in the pipeline increases the quality of obtained models if compared to the other tested releases and obtained significantly lower computational costs, which is always interesting in a distributed system running on a central server as AutoModel.

Table 1. Comparison of different versions of AutoModel with ModWeb and Swiss-Model when modeling the vicilin, TvLDH and gi | 157108525 | sequences.

Software	AutoModel 0.4 (9v4)	AutoModel 0.4 (9v9)	AutoModel 0.5 (HMMER)	AutoModel 0.5 (Muscle)	ModWeb	Swiss-Model
Sequence 1	7S Vicilin from *A. angustifolia* [GenBank: AAM81249.1]					
Template	1uijA	1uijA	1uijA	1uijA	2ea7A	1uijA
Model qual. (Z-score)	−4,85	−4,65	−1,21	−4,41	−4,48	−5,43
Time lapse of alignment	2 m 18 s	2 m 07 s	1 m 27 s	1 s	-	-
Full Modeling time	7 m 44 s	8 m 10 s	7 m	6 m 14 s	10 h ± 1 h	1 h 10 m ± 10 m
Sequence 2	Lactate dehydrogenase from *T. vaginalis* (TvLDH) [GenBank: AF070994]					
Template	1bdmA	1bdmA	1bdmA	1bdmA	4uulA	1bdmA
Model qual. (Z-score)	−8,9	−8,43	−5,23	−8,18	−10,43	−9,05
Time lapse of alignment	1 m 05 s	1 m 21 s	1 m 40 s	1 s	-	-
Full Modeling time	6 m 50 s	5 m 41 s	7 m 20 s	5 m 28 s	7 h ± 5 h	45 min ± 5 m
Sequence 3	Developmental Protein Cactus From *A. aegypti* [refseq:XP_001650267]					
Template	1iknD	1iknD	1iknD	1iknD	1k1aA	5leaA
Model qual. (Z-score)	5,26	−1,44	−0,76	−2,66	−6,36	−4,42
Time lapse of alignment	32 s	52 s	2 m 10 s	1 s	-	-
Full Modeling time	4 m 22 s	5 m	6 m	3 m 48 s	13 h ± 2 h	1 h 13 m ± 5 m
Other structural information from the template						
Model with crystallographic water?	Yes				No	No
Model with ligands?	Yes				No	Yes

4   Conclusions

The AutoModel was developed to have an easy to use graphical interface to guide new and non-expert user in homology modeling experimental pipeline. Unlike other implemented modeling tools, AutoModel allows the user to control all the experimental steps of the modeling pipeline, even including ligands and structural waters from the structural template to the modeled protein. Although the last tested release of Modeller (9v9) improved the speed of the alignment procedure and the quality of the obtained results, a problem detected in the 9v4 release of the program, the use of Muscle as alignment tool in the Automodel pipeline increased the quality of obtained models and obtained significantly lower computational costs, important characteristic in a distributed system running on a central server. The main goal of AutoModel is to provide a friendly interface to allow unexperienced user to control several parameters of experiment in client-server architecture. In the presented release, the AutoModel 0.5 allows the user to select template, The main of AutoModel is to provide a friendly interface to allow unexperienced user to control several parameters of experiment in client-server architecture. In the presented release, the AutoModel 0.5 allows the user to select template, select heteroatoms, import structural information of presented ligand from the template, and even refine specific loops in a quick and easy way without requiring expensive hardware or advanced knowledge in Computational Biology.

4.1 Availability and Requirements

The AutoModel Client is distributed "*as is*" under GNU/GPL license, and may be considered as freeware for scientific, academic and student users. The client of AutoModel is available to download in http://biocomp.uenf.br. In this page potential users will obtain introductory manual, general information of the project and access to the AutoModel Client download page. On the other hand, "AutoModel Server" and "AutoModel Client" packages are available for Linux users through pypi package index. Thus, with few commands, as ">pip install automodel-server", the AutoModel and its python dependencies are automatically installed in Linux servers. Beside this, AutoModel server performs template downloads automatically from PDB website, thus making obsolete the needs of "on site" bulky databases. The AutoModel Client and Server are available for academic community "as is" for free in http://biocomp.uenf.br.

In order to use AutoModel Client, a netbook with Linux operating system and internet access with minimum of 2 GB of RAM will be sufficient hardware but authors recommend Ubuntu or Mint Linux distributions and 4 GB RAM laptop. Furthermore, the full functional AutoModel Client requires a third part WxPython [27] and RPyC libraries, and Python program installed in your system. For protein structure visualization, Pymol [34] and/or VMD [35] must be also installed in your computer. All the information necessary for installing these dependencies is available in our website.

Acknowledgements. This research was supported by E-26/110.216/2011 FAPERJ grant for J. H.F. and FAPERJ master degree grant for J.de A.F.

References

1. Rodwell, V., Bender, D., Botham, K.M., Kennelly, P.J., Weil, P.A.: Harpers Illustrated Biochemistry, 30th edn. McGraw Hill Professional, New York (2015)
2. Hillisch, A., Pineda, L.F., Hilgenfeld, R.: Utility of homology models in the drug discovery process. Drug Discov. Today **9**, 659–669 (2004)
3. Sliwoski, G., Kothiwale, S., Meiler, J., Lowe, E.W.: Computational methods in drug discovery. Pharmacol. Rev. **66**, 334–395 (2014)
4. Schmidt, T., Bergner, A., Schwede, T.: Modelling three-dimensional protein structures for applications in drug design. Drug Discov. Today **19**, 890–897 (2014)
5. Tang, M., et al.: High-resolution membrane protein structure by joint calculations with solid-state NMR and X-ray experimental data. J. Biomol. NMR **51**, 227–233 (2011)
6. Krieger, E., Nabuurs, S.B., Vriend, G.: Homology modeling. Methods Biochem. Anal. **44**, 509–524 (2003)
7. Zhang, Y.: Interplay of I-TASSER and QUARK for template-based and ab initio protein structure prediction in CASP10. Proteins Struct. Funct. Bioinform. **82**, 175–187 (2014)

8. Roy, A., Kucukural, A., Zhang, Y.: I-TASSER: a unified platform for automated protein structure and function prediction. Nat. Protoc. **5**, 725–738 (2010)
9. RosettaCommons. https://www.rosettacommons.org/
10. CASP12. http://www.predictioncenter.org/casp12/
11. Conchúir, S.Ó., et al.: A web resource for standardized benchmark datasets, metrics, and Rosetta protocols for macromolecular modeling and design. PLoS ONE **10**, e0130433 (2015)
12. Webb, B., Sali, A.: Protein structure modeling with MODELLER. In: Kaufmann, M., Klinger, C., Savelsbergh, A. (eds.) Functional Genomics, vol. 1654, pp. 39–54. Springer, Heidelberg (2017)
13. Eswar, N., Webb, B., Marti-Renom, M.A., Madhusudhan, M.S., Eramian, D., Shen, M., Pieper, U., Sali, A.: Comparative protein structure modeling using Modeller. Curr. Protoc. Bioinform. **15**, 5–6 (2006)
14. Webb, B., Sali, A.: Comparative protein structure modeling using Modeller. Curr. Protoc. Bioinform. **47**, 5–6 (2014)
15. Finn, R.D., Clements, J., Eddy, S.R.: HMMER web server: interactive sequence similarity searching. Nucl. Acids Res. **39**, gkr367 (2011)
16. Larkin, M.A., et al.: Clustal W and Clustal X version 2.0. Bioinformatics **23**, 2947–2948 (2007)
17. Laskowski, R.A., MacArthur, M.W., Moss, D.S., Thornton, J.M.: PROCHECK: a program to check the stereochemical quality of protein structures. J. Appl. Crystallogr. **26**, 283–291 (1993)
18. Schrödinger. https://www.schrodinger.com/. Accessed 19 Aug 2018
19. Certara – Certara is the leading drug development consultancy with solutions spanning the discovery, preclinical and clinical stages of drug development. https://www.certara.com/
20. Schwede, T., Kopp, J., Guex, N., Peitsch, M.C.: SWISS-MODEL: an automated protein homology-modeling server. Nucleic Acids Res. **31**, 3381–3385 (2003)
21. MHOLline. http://www.mholline.lncc.br/
22. Lambert, C., Léonard, N., De Bolle, X., Depiereux, E.: ESyPred3D: prediction of proteins 3D structures. Bioinformatics **18**, 1250–1256 (2002)
23. Fiser, A., Do, R.K.G., Šali, A.: Modeling of loops in protein structures. Protein Sci. **9**, 1753–1773 (2000)
24. Marti-Renom, M.A., Madhusudhan, M.S., Sali, A.: Alignment of protein sequences by their profiles. Protein Sci. **13**, 1071–1087 (2004)
25. Cock, P.J.A., et al.: Biopython: freely available Python tools for computational molecular biology and bioinformatics. Bioinformatics **25**, 1422–1423 (2009)
26. RPyC - Transparent, Symmetric Distributed Computing. https://rpyc.readthedocs.io/en/latest/
27. Talbot, H.: wxPython, a GUI Toolkit. Linux J. **2000**, 5 (2000)
28. Shen, M.: Statistical potential for assessment and prediction of protein structures. Protein Sci. **15**, 2507–2524 (2006)
29. Kelly, C.A., Nishiyama, M., Ohnishi, Y., Beppu, T., Birktoft, J.J.: Determinants of protein thermostability observed in the 1.9-. ANG. crystal structure of malate dehydrogenase from the thermophilic bacterium Thermus flavus. Biochemistry **32**, 3913–3922 (1993)
30. Wu, G., Fiser, A., Ter Kuile, B., Šali, A., Müller, M.: Convergent evolution of Trichomonas vaginalis lactate dehydrogenase from malate dehydrogenase. Proc. Natl. Acad. Sci. **96**, 6285–6290 (1999)
31. Šali, A.: Tutorial (2008)

32. Maruyama, N., Maruyama, Y., Tsuruki, T., Okuda, E., Yoshikawa, M., Utsumi, S.: Creation of soybean β-conglycinin β with strong phagocytosis-stimulating activity. Biochim Biophys Acta (BBA)-Proteins. Proteomics **1648**, 99–104 (2003)
33. Wiederstein, M., Sippl, M.J.: ProSA-web: interactive web service for the recognition of errors in three-dimensional structures of proteins. Nucl. Acids Res. **35**, W407–W410 (2007)
34. DeLano, W.L.: The PyMOL molecular graphics system (2002)
35. Humphrey, W., Dalke, A., Schulten, K.: VMD: visual molecular dynamics. J. Mol. Graph. **14**, 33–38 (1996)

Detecting Acute Lymphoblastic Leukemia in down Syndrome Patients Using Convolutional Neural Networks on Preprocessed Mutated Datasets

Maram Shouman[✉], Nahla Belal, and Yasser El Sonbaty

Arab Academy for Science and Technology,
College of Computing and Information Technology, Computer Science,
Alexandria, Egypt
maram.shouman18@yahoo.com, {nahlabelal,yasser}@aast.edu

Abstract. Convolutional neural networks extract high-level abstraction features using minimum preprocessing steps. In this research, we propose a new approach in classifying Down Syndrome with Acute Lymphoblastic Leukemia using a convolutional neural network. Sequences are represented using a one hot vector depending on point mutation as input to the CNN model. Therefore, it conserves the necessary position data of each nucleotide in the sequence. Using two different genomic datasets, our proposed model has achieved significant improvements over classical classification techniques, with an increased accuracy of 98%, and 98.5%, respectively.

Keywords: Down Syndrome · Mutation Detection Techniques
Convolutional neural network

1 Introduction

The complete set of DNA that is used to build up the organism is called a genome. Sequencing the genome is an important step in understanding it. Also, the genome contains information about where genes are. Studying the entire genome sequence helps in understanding the whole work of the genomes, how they direct the growth, maintenance, and development of the organism [24]. A genetic disease [13] is caused by some mutation or change in the DNA. Any variation in a DNA sequence is known as mutation. Genes are vital for human lives, as they code for the protein that builds up the cell structures and carry out most of life's functionality. When a gene is mutated, its protein product can never again complete its typical functionality, causing a disorder disease [15].

Down Syndrome is an example of a disorder that is caused by the presence of an extra chromosome. Patients with Down syndrome (DS) have a high probability to get acute myeloid leukemia (AML) and acute lymphoblastic leukemia

© Springer Nature Switzerland AG 2018
R. Alves (Ed.): BSB 2018, LNBI 11228, pp. 90–102, 2018.
https://doi.org/10.1007/978-3-030-01722-4_9

(ALL), due to the mutation caused in their DNA. Also, they are characterized by specific biological features in contrast with non DS-ALL [3].

Cancer is a prime example of heterogeneous disease. Acute lymphocytic leukemia (ALL) [14] is a type of blood cancer, that is also known as acute lymphoblastic leukemia. Moreover, it starts from white blood cells called lymphocytes in the bone marrow. Leukemia cells generally attack the blood fast. They would then be able to spread to different parts of the body, including the lymph nodes, bone, and central nervous system. The expression "acute" implies that the leukemia can grow rapidly, and if not treated, will most likely cause death in a couple of months.

Nowadays, it is easy to read a genome sequence due to the rapid development of sequencing technologies. Moreover, the databases in Genbank and NCBI have grown rapidly in the last few years. Motivated by the importance of the biological problem, low accuracy obtained by regular classification techniques [28], as well as the availability of huge databases, modern machine learning techniques, and especially deep learning techniques which can be applied to help in understanding the genome and identifying diseases [8], we developed a new encoding technique to be able to present the mutated genes in CNN architecture, which shows impressive results.

Deep learning [7] is a new branch of machine learning techniques that was introduced in the last decades. It contains different types of models in which they have multiple non-linear transforming layers that extract features with a high level of abstraction. It was able to show impressive results [23] and to solve very complicated problems. Deep learning techniques, especially convolutional neural networks have shown great improvement in different fields such as speech recognition, image processing [7], 3d segmentation, and computer vision [23].

Convolutional neural networks (CNNs) is a specific type of deep learning algorithms to overcome the problem in traditional machine learning algorithms which requires manual feature extraction before the classification process. CNNs not only perform classification, but they can also learn to extract features directly [7]. CNN provides the flexibility of extracting intrinsic and discriminating features from genomes, that are most suitable for classification [20].

CNN accepts data in numeric form only. CNN is powerful in solving complex and large-scale problems in different fields [5,16,26]. In this research, a deep learning model using a one-hot vector is proposed to represent sequences based on their mutation for detecting genetic disease. This model was inspired by a deep learning model for text classification [20].

Our research applies deep learning techniques to capture mutated sequences with the disease, and the hidden semantics and interconnections. In this research, we apply a CNN deep learning model and compare it with classical machine learning techniques such as Support Vector Machines (SVM), and k-means as well as seq-CNN in [20], and one hot encoding method in [4]. Moreover, we present a new conversion method for mutated sequences as input for CNN.

The paper is organized as follows. Section 2 presents the related work. Our proposed model is presented in Sect. 3. Section 4 gives the experimental results and evaluation. Finally, Sect. 5 gives the conclusion and future work.

2 Related Work

One of the deep learning models that applies convolutional layers in extracting features from the data is the convolutional neural network model. In this model, the extracted features from the previous layer are used by neurons in a convolutional layer to extract higher-level features.

One of the applications of convolutional neural networks is face recognition [9]. CNNs can effectively model multidimensional data, and are shown to be powerful in solving computer vision [23], and image recognition problems [26]. Although there are some researchers that applied CNN to solve biological gene expression regulation sequences problems [18], the depth of CNN model allows the applicability of complex learning patterns as well as identifying longer motifs, and sophisticated regulatory codes [21].

To employ the CNN for solving the DNA motif discovery problem, existing works typically encode nucleotide in DNA sequences using the one-hot vector method [4,19]. That is, each base pair in a sequence is encoded with a binary vector of four bits with one of it is hot (i.e. 1) while others are 0. For instance $A = (1, 0, 0, 0)$, $G = (0, 1, 0, 0)$, $C = (0, 0, 1, 0)$, and $T = (0, 0, 0, 1)$. This sequence representation method draws the similarity to the Position Frequency Matrix in [29], in which a vector indicates the probability of occurrences of the four bases at a certain position in a DNA sequence. Therefore, an input DNA sequence of length L is represented as $4 \times L$ matrix, where L is the length of the sequence.

Moreover, several researchers have investigated converting biological sequences into numerical values [2,4,29]. These methods can be divided into direct and indirect encoding [2]. Direct encoding methods use a numerical value for each nucleotide or they use a vector of numerical values for each nucleotide. However, the indirect methods apply a set of features from the biological sequences. The features can be dependent on the frequency counts of k-mers, biological, or biochemical properties. For instance, in [4] A is represented by 0.25, C by 0.50, G by 0.75, and T by 1.00, it is difficult to justify how those numbers are decided rather than heuristic. The evaluation study compares three sequence representation methods: (a) one-hot vector; (b) ordinal encoding with square matrix (square), and (c) 1D vector. It shows an accuracy of 87%, 92.6%, and 91.2%, respectively.

In [29], the convolutional neural network architecture takes its input as a $4 \times L$ matrix where L is the length of the sequence. Each base pair in the sequence is denoted as one of the four one-hot vectors $[1, 0, 0, 0]$, $[0, 1, 0, 0]$, $[0, 0, 1, 0]$ and $[0, 0, 0, 1]$ and gives an accuracy of 80%. While in [23], after applying Deep Cons convolutional neural network which involved one input layer, with three hidden layers, and one output layer. The accuracy improved to 83%.

The DNA sequences consist of successive letters without space where there is no term of a word in it. Text recognition is similar to this problem, where they both contain letters, one followed by spaces and the other not. Researchers applied convolutional neural network text data problems such as topic categorization, spam detection, and sentiment classification. As Convolutional neural networks take only numeric data, it is a challenge to convert these data into numeric form without losing any information. Moreover, unlike image recognition problems, that is mapped into two-dimensional numerical matrices, text data are one-dimensional sequences of successive letters without spaces. Some researchers solved this problem by generating a lookup table with a represented vector that is used to match each word in the vocabulary [17]. However, some researchers [6] claimed using a lookup table, another approach is proposed to represent words using one hot vector. Moreover, they concatenated n-gram information and used the representation as an input to convolutional neural networks. This model showed outperformance in classification problems. In [20], a method to translate DNA sequences to a sequence of words, in order to apply the same representation technique for text data in CNN without losing position information of each nucleotide in sequences is proposed.

Mutations are important information to detect some diseases. Different kinds of mutation are classified in [13], with the most observed types in dominant disorders being gain of function, haploinsufficiency, and dominant negative. The gain-of-function mutations cause an increase in the amount of gene product or its activity, and sometimes create a new product, leading to a toxic product responsible for a pathological effect. Dominant negative mutations results from change in the molecular function, which may cause implications of cancer, while Haploinsufficiency can occur through different ways. First, the production message may be erased due to mutation in the gene. Second, some parts of the gene are missing due to deletion. Finally, ins of the protein.

Point Mutation is a single base pair alteration; it includes a transition from one nucleotide to another. Many mutation detection techniques have been derived in the last few decades [1,11,25]. In [25] a good comparison is made between five different somatic point mutation detection techniques (JointSNVMix, MuTect, Somatic Sniper, Strelka, and Varscan2). Strand-biased are not highly detected in many tools such as Strelka, especially of low quality. However, in MuTect and Varscan 2 the user decides whether to detect strand-biased or not. According to the results of the experiment, MuTect showed better results than the other tools especially in characterizing low allelic fraction SNV (Single nucleotide variation). Similarly in [1] a comparison is held between six different tools (Deep-SNVMiner, FreeBayes, GATK, LoFreq, and SAMTools). It is compared with different dilution (0, 90, 99, 99.9, 99.99, 99.999, 99.9999%), Deep-SNVMiner detected different mutation till 99.9%, LoFreq detected mutation till 99%, GATK detected mutation till 90%, while SAMTools and FreeBayes detected at level 0% only. Also in [10,22] provided a very useful implementation for Samtools adding more functionality.

3   Proposed Model

In this paper, we present a deep learning method for studying mutated sequences. Moreover, a certain threshold is required to identify DS with Leukemia, this threshold is calculated as the sum of both mutated nucleotide (C to T) and unmutated nucleotide (C). The convolutional neural network is used to predict whether a Down syndrome patient has leukemia or not. By learning to discriminate between mutated and un-mutated nucleotides to be represented as a binary code. This binary code is then used as input for CNN.

The main contribution of this research is the process of converting the values of the DNA bases to numeric values, while maintaining the information about the mutation that occurred. CNN accepts data in numeric form only, the proposed model preprocesses the data to convert the DNA mutated sequence and un-mutated data into numeric values without losing any information.

The proposed algorithm takes as input a mutated dataset. The first step is to detect the single nucleotide polymorphism. Afterward, convert it to numeric form, for example, if the original genome is A and has not been changed, then the binary code will be 0001. While if it is mutated to C it will be 0010, but if it is mutated to T it will be 0100, as illustrated in Table 2. These binary codes are used as input to CNN. The following block diagram explains the structure of our algorithm. Moreover, each block is illustrated in details in the following subsections (Fig. 1) and Table 1.

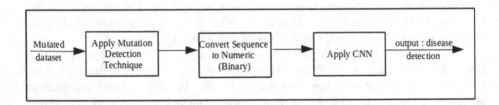

Fig. 1. Proposed model block diagram.

A. Apply Mutation Detection Technique:

In this paper, Samtools Software is used to detect the mutation based on multiple reasons illustrated in the following paragraph.

The SAMtools utilities comprise a very helpful and widely used suite of the software system for manipulating files and alignments within the sam format. The large SAM files will be regenerated to the binary equivalent BAM files. Different implementation of samtools are available in different programming languages such as in Perl, Python, and Java. Availability: http://samtools. sourceforge.net

Moreover, samtools depends on the second error which comes at the next likelihood in SNP genotype. One of the main difference between samtools and the othe software is Indel genotype likelihood model. Samtools' model springs from BAQ. In order to identify alleles it uses some ad hoc heuristics. Moreover, it employes hand tuned filters and collects statistics about the alignment to guarantee effective filtering. Also it is scalable for big data with a simple framework. Moreover, it is provided with specific features such as genotype-free analysis, and physical phasing. In addition, samtools concentrate on using filters to reduce bias and produce consensus in the right way.

B. Convert Sequence to Numeric (Binary) form: DNA sequences are formed from successive letters without any spaces and the mutated sequences may cause different diseases. Therefore, a model is proposed in order to represent these data without losing any information of each nucleotide in the sequences. DNA Sequences consist of four values A, G, C, and T. With regular datasets, it needs only two bits to represent these four values. However, mutation plays a critical role in genetic disease detection. Specifically as mentioned [12], in DS with Leukemia mutation from C to T is required to identify the disease. There are two cases for each nucleotide. The first case is mutated nucleotide such as from A to C, A to G, or A to T. Similarly, the remaining nucleotides will have their possible mutations. While un-mutated nucleotides remain unchanged. Un-mutated nucleotides sum up to four possibilities. While mutated nucleotides give 12 possibilities (each nucleotide can be mutated to the other 3.). Hence, the total number of possibilities is 16, which needs four bits to represent. Table 2, illustrates the discussed representation. Each possible mutation is represented using four binary digits. In table 2, the dot notation represents no mutation. In row 1, if A is detected with no mutation, it will be encoded to 0001. While in row number 8 a mutation detected from C to T will be encoded to 1000.

C. Apply CNN:
In this paper, the DNA sequence is divided into a window size of three, with one slide width as in [29] which obtains the best results, each nucleotide is mapped to a numeric value depending on whether it is mutated or not, as illustrated in the previous table. Every three successive nucleotides are treated as a word then converted to a 2d array to be used as an input to the CNN.
Figure 2 shows an example of a DNA sequence that is mapped according to the mutation detected in each nucleotide. As shown in Fig. 2, if A remains unchanged the code will be 0001, while C is mutated to T so its code will be 1000, and G is mutated to A has the code 1011. These three nucleotides determine the first word. Afterward the window will slide one step, so the next word will contain CGT. This process continues until the end of the sequence. Then every two words will form the vertical data for the 2D matrix. This matrix is used as an input for CNN.
The convolutional neural network depends on certain steps as shown in Fig. 3. The primary step, is the layer that receives the input and tries to label the input by bearing on what it has learned within the past. The resulting output is then passed on to the subsequent layer. Intuitively, every convolution filter represents a feature of interest, and also the CNN algorithm learns which features comprise a specific class. The output strength depends on the presence or absence of a specific feature and not on its location. The second Step is

Subsampling, that aims to scale back the sensitivity of the filters to noise and variations. This could be achieved by taking averages of the input. The third step is the activation layer that controls how the signal flows from one layer to the following. Output signals which are powerfully related to past references would activate additional neurons, enabling signals to be propagated more expeditiously for identification. The fourth step is the fully Connected, where the last layers through the network are all connected, In other words, all neurons of preceding layers are connected to every neuron in consequent layers. This mimics high-level reasoning where all possible pathways from the input to output are considered. Moreover, once training the neural network, there is an extra layer known as the loss layer. This layer provides feedback to the neural network on whether or not it identified inputs properly, and if not, how far its guesses were. This helps to guide the neural network to strengthen the correct concepts as it trains. The proposed model contains six convolutional layers. Each of these layers is followed by two subsampling layers which are used to extract features from sequences matrices. Then apply a fully connected neural network with 50 neurons and to decrease the effect of overfitting we used a dropout value of 0.5. Finally, the softmax output layer is used to determine whether the sequence contains leukemia or not. In which this Configuration shows the best results.

Table 1. DNA nucleotide mutation binary code.

Model			
No	DNA nucleotide	Mutated nucleotide	Binary code
1	A	.	0 0 0 1
2	C	.	0 1 0 1
3	G	.	1 0 0 1
4	T	.	1 1 0 1
5	A	C	0 0 1 0
6	A	T	0 1 0 0
7	A	G	0 0 1 1
8	C	T	1 0 0 0
9	C	A	0 1 1 0
10	C	G	0 1 1 1
11	G	A	1 0 1 1
12	G	C	1 1 0 0
13	G	T	1 0 1 0
14	T	G	0 0 0 0
15	T	A	1 1 1 0
16	T	C	1 1 1 1

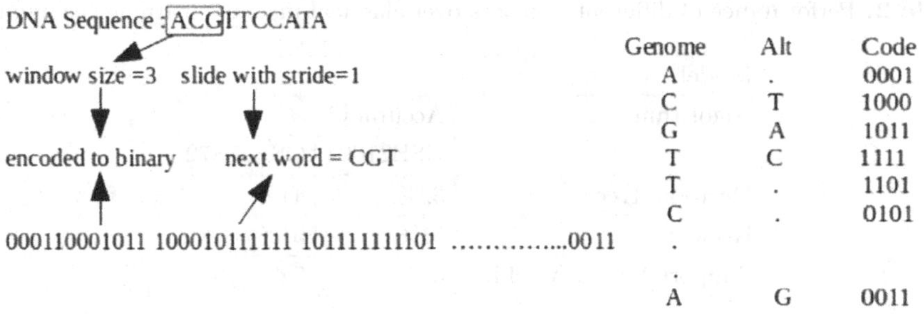

Fig. 2. Representation of DNA sequence in CNN

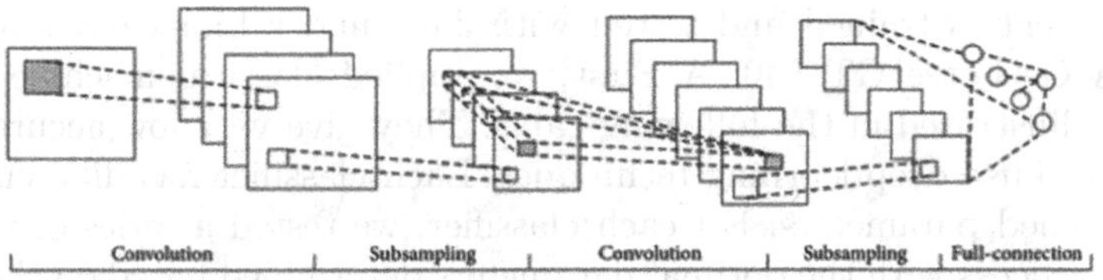

Fig. 3. Architecture of convolutional neural network. [27]

4 Experiment and Results

Dataset:
In order to evaluate the performance of the proposed model in solving point mutation disease detection problem, and to facilitate evaluation and comparison, different widely used and publicly available ALL-DS and ALL-NDS [12] are used. Datasets are available on NCBI with accession number: GSE21094. This dataset was released in 2011 and updated in 2017, Moreover, it contains genome-wide profiling of 58 DS-ALL and 35 non-Down syndromes (NDS) ALL. Each sample holds about 27578 records. Only 2 datasets are used in this experiment as the rest do not contain genomic data. For each sample, 27578 records are used, and 8274 records are randomly selected as the testing set, and the remaining are taken as the training set. The first is provided with IlmnID, Name, IlmnStrand, SNP, AddressA ID, AlleleA ProbeSeq, AddressB ID, AlleleB ProbeSeq, GenomeBuild, Chr, MapInfo, Ploidy, Species, Source, SourceVersion, SourceStrand, SourceSeq, TopGenomicSeq, BeadSetID, Intensity Only, Exp Clusters, CNV Probe. The second is provided with AddressA ID, AlleleA ProbeSeq, AddressB ID, AlleleB ProbeSeq, GenomeBuild, Chr, MapInfo, Ploidy, Species, Source, SourceVersion, SourceStrand, SourceSeq, TopGenomicSeq, BeadSetID, Intensity Only, Exp Clusters for more details please see [12]. These datasets include down sydrome with leukemia, and without leukemia. Moreover, it is provided with an average beta value which is calculated as the percent signal from a methylated probe (C not converted to T) proportional to the sum of both methylated (C not converted to T) and unmethylated probes (C bisulfite-converted to T) (value = C/[(T + C) + 100]) [12].

Table 2. Performance of different datasets over classical machine learning techniques.

Model		
Algorithm	Accuracy	
	GSE21091	GSE20872
Decision Tree	32%	41%
K-means	51%	51%
Support Vector Machine	39%	52%

Experimental Configuration:

The experiment is trained and tested with Java under Linux environment on a NVIDIA GetForce GT 540. A First, we applied classical machine learning techniques illustrated in the following table. They give very low accuracy, then we decided to use deep learning techniques. Each classifier has different tuning steps and tuned parameters. For each classifier, we tested a series of values for the tuning process with the optimal parameters determined based on the highest overall classification accuracy.

In SVM, the radial basis function (RBF) kernel of the SVM classifier is often used and shows the most effective performance. Therefore, we tend to used the RBF kernel to implement the SVM algorithm. There are two parameters that need to be set once applying the SVM classifier with RBF kernel: the optimum parameters of the kernel width parameter (Y) and also the cost (C) it had been choosen to administer the most effective results.

In Decision tree, the default values for the parameters controlling the size of the trees (e.g. `max_depth`, `min_sample_leaf`) result in growing and unpruned trees which may potentially be very large on some data sets. To scale back memory consumption, the complexity and size of the trees is controlled by setting those parameter values to provide the most effective results.

Then applied K-Means clustering with different conventional initial centroid selection methods and gets the higher accuracy with $K = 2$.

DeepLearnToolbox is used which provides an implementation for text categorization on GPU using a convolutional neural network to implement the model. The model contains 6 convolutional layers. Each of these layers is followed by 2 sub-sampling layers. These layers are used to extract features from the samples with a weight of 0.001. For CNN training, we use mini batches of 50 training samples. Other hyperparameters of the model were chosen based on its performances in datasets for tuning.

Experimental Results:

After applying the classical machine learning techniques that has shown low accuracy, we decided to apply our proposed model and compare it with the one hot vector and Sqaure matrix [29] and [4]. One hot vector that represent the $A = (1, 0, 0, 0)$, $G = (0, 1, 0, 0)$, $C = (0, 0, 1, 0)$ and $T = (0, 0, 0, 1)$ while, the Square matrix represent the A by 0.25, C by 0.50, G by 0.75 and T by 1.00. Then apply CNN (Table 3).

Table 3. Performance of different datasets over different CNN.

Model		
Algorithm	Accuracy	
	GSE21091	GSE20872
CNN	98%	98.5%
One hot vector [29]	79%	79.5%
Square matrix [4]	80%	81%

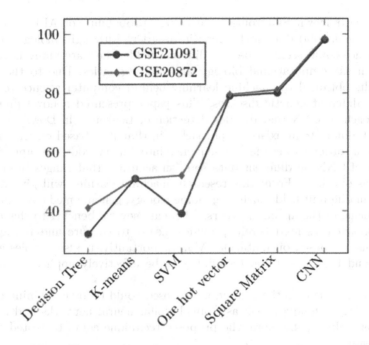

Fig. 4. Accuracy of different machine learning techniques.

The accuracy is calculated as the percentage of correctly classified instances $(TP + TN)/(TP + TN + FP + FN)$. Where TP, FN, FP and TN represent the number of true positives, false negatives, false positives and true negatives, respectively (Fig. 4).

As evidenced in the previous diagram and table, our proposed model system shows higher accuracies than the system proposed in [29] and [4], with an improvement of 19% and 18%, respectively. Furthermore, it shows that the proposed model is more robust in detecting leukemia in Down Syndrome patients. Moreover, the proposed end-to-end disease detection system did not require an extremely specialized lexicon to achieve high performance, this is due to the identification of mutations that cause the disease. Compared to the ordinary hot vector classifier in [29], and square matrix in [4], which showed 79% and 80% accuracy, respectively, our proposed disease detection classifier showed an accuracy of almost 98%. Moreover, the increase of the accuracy of the mutation detection binary code will enhance the ability to discern leukemia, thus, form better estimates to identify any genetic disease. This result shows that by constructing a high-performing disease recognition system, a state-of-the-art genetic disease recognition can be obtained without needing to leverage more classical machine learning methods such as SVM, Kmeans, and Decision Table.

5   Conclusion and Future Work

Children with Down Syndrome have a high probability of ALL (acute lymphoblastic leukemia) due to the genetic disorder. Although, biology and computer science could seem to be two different fields, their are many intersection fields such as computational biology and bioinformatics. Due to the impressive results obtained by machine learning behind computer science to help in detecting different genetic diseases. This paper presented a novel preprocessing approach to CNN that enables detection of Leukemia in Down Syndrome. This work differs from existing approaches in that it is based on the mutation detection in order to decode the sequence into binary code. By applying this approach of CNN on different data sets, an accuracy that ranges from 98% to 98.5% was obtained. From this research, machine learning will give dramatic progress in different fields such as genome biology, genome medicine, and precision medicine in the upcoming years. There are several benefits to this method. The approach presented herein provides means to capture and represent the genetic disease based on mutation. More importantly, it also enables efficient, existing and future, solution techniques to be effectively applied to any SNP disease.

An enhancement to the technique proposed could be by combining different deep learning techniques such as Convolutional neural networks and recurrent neural network. Furthermore, the proposed technique could be tested on more SNP diseases.

References

1. Andrews, T.D., Jeelall, Y., Talaulikar, D., Goodnow, C.C., Field, M.A.: Deep-SNVMiner: a sequence analysis tool to detect emergent, rare mutations in subsets of cell populations. PeerJ **4**, 1–13 (2016)
2. Blekas, K., Fotiadis, D.I., Likas, A.: Motif-based protein sequence classification using neural networks. J. Comput. Biol. **12**(1), 64–82 (2005)
3. Buitenkamp, T.D., et al.: Acute lymphoblastic leukemia in children with down syndrome: a retrospective analysis from the ponte di legno study group. Blood **123**(1), 70–77 (2014)
4. Choong, A.C.H., Lee, N.K.: Evaluation of convolutionary neural networks modeling of DNA sequences using ordinal versus one-hot encoding method. bioRxiv, pp. 60–65 (2017)
5. Hannun, A., et al.: Deep speech: scaling up end-to-end speech recognition. arXiv preprint arXiv:1412.5567 (2014)

6. Johnson, R., Zhang, T.: Effective use of word order for text categorization with convolutional neural networks. arXiv preprint arXiv:1412.1058 (2014)
7. Krizhevsky, A., Sutskever, I., Hinton, G.E.: ImageNet classification with deep convolutional neural networks. In: Advances in Neural Information Processing Systems, pp. 1097–1105 (2012)
8. Leung, M.K.K., Delong, A., Alipanahi, B., Frey, B.J.: Machine learning in genomic medicine: a review of computational problems and data sets. Proc. IEEE **104**(1), 176–197 (2016)
9. Li, H., Lin, Z., Shen, X., Brandt, J., Hua, G.: A convolutional neural network cascade for face detection. In: Proceedings of the IEEE Conference on Computer Vision and Pattern Recognition, pp. 5325–5334 (2015)
10. Li, H.: A statistical framework for SNP calling, mutation discovery, association mapping and population genetical parameter estimation from sequencing data. Bioinformatics **27**(21), 2987–2993 (2011)
11. Li, H., et al.: The sequence alignment/map format and samtools. Bioinformatics **25**(16), 2078–2079 (2009)
12. Loudin, M.G., et al.: Genomic profiling in down syndrome acute lymphoblastic leukemia identifies histone gene deletions associated with altered methylation profiles. Leukemia **25**(10), 1555 (2011)
13. Mahdieh, N., Rabbani, B.: An overview of mutation detection methods in genetic disorders. Iran. J. Pediatr. **23**(4), 375–388 (2013)
14. Maloney, K.W.: Acute lymphoblastic leukaemia in children with down syndrome: an updated review. Br. J. Haematol. **155**(4), 420–425 (2011)
15. McCarthy, M.I., MacArthur, D.G.: Human disease genomics: from variants to biology. Genome Biol. **18**(1), 1–3 (2017)
16. Mikolov, T.: Statistical language models based on neural networks. Presentation at Google, Mountain View (2012)
17. Mikolov, T., Chen, K., Corrado, G., Dean, J.: Efficient estimation of word representations in vector space. arXiv preprint arXiv:1301.3781 (2013)
18. Min, S., Lee, B., Yoon, S.: Deep learning in bioinformatics. Brief. Bioinform. **18**(5), 851–869 (2017)
19. Ng, P.: dna2vec: Consistent vector representations of variable-length k-mers. arXiv preprint arXiv:1701.06279 (2017)
20. Nguyen, N.G., et al.: DNA sequence classification by convolutional neural network. J. Biomed. Sci. Eng. **9**(05), 280–286 (2016)
21. Pan, X., Shen, H.-B.: Predicting RNA-protein binding sites and motifs through combining local and global deep convolutional neural networks. Bioinformatics p. bty364 (2018)
22. Ramirez-Gonzalez, R.H., Bonnal, R., Caccamo, M., MacLean, D.: Bio-samtools: ruby bindings for samtools, a library for accessing bam files containing high-throughput sequence alignments. Source Code Biol. Med. **7**(1), 1–6 (2012)
23. Srinivas, S., et al.: A taxonomy of deep convolutional neural nets for computer vision. arXiv preprint arXiv:1601.06615 (2016)
24. Venter, J.C., et al.: The sequence of the human genome. Science **291**(5507), 1304–1351 (2001)
25. Wang, Q., et al.: Detecting somatic point mutations in cancer genome sequencing data: a comparison of mutation callers. Genome Med. **5**(10), 1–8 (2013)

26. Wu, R., Yan, S., Shan, Y., Dang, Q., Sun, G.: Deep image: scaling up image recognition, **7**(8). arXiv preprint arXiv:1501.02876 (2015)
27. Yan, S., Xia, Y., Smith, J.S., Lu, W., Zhang, B.: Multiscale convolutional neural networks for hand detection. Appl. Comput. Intell. Soft Comput. **2017** (2017)
28. Yue, T., Wang, H.: Deep learning for genomics: a concise overview. arXiv preprint arXiv:1802.00810 (2018)
29. Zeng, H., Edwards, M.D., Liu, G., Gifford, D.K.: Convolutional neural network architectures for predicting DNA-protein binding. Bioinformatics **32**(12), i121–i127 (2016)

S^2FS: Single Score Feature Selection Applied to the Problem of Distinguishing Long Non-coding RNAs from Protein Coding Transcripts

Bruno C. Kümmel[1](✉), Andre C. P. L. F. de Carvalho[2], Marcelo M. Brigido[3], Célia G. Ralha[1], and Maria Emilia M. T. Walter[1]

[1] Department of Computer Science, University of Brasilia, Brasilia, DF, Brazil
bruno.kummel@aluno.unb.br, {ghedini,mariaemilia}@unb.br
[2] Department of Computer Sciences, University of Sao Paulo, Sao Carlos, SP, Brazil
andre@icmc.usp.br
[3] Department of Cellular Biology, University of Brasilia, Brasilia, DF, Brazil
brigido@unb.br

Abstract. The task of distinguishing long non-coding RNAs (lncRNAs) from protein coding transcripts (PCTs) has been previously addressed with machine learning (ML) algorithms using hundreds of features. However, the use of a large number of features can negatively affect the predictive performance of these algorithms since it can lead to problems like overfitting due to a phenomenon known as the curse of dimensionality. In order to deal with these problems, dimensionality reduction techniques have been proposed, among them, feature selection. This work proposes and experimentally evaluates a simple and fast feature selection technique, called Single Score Feature Selection - S^2FS.

For such, initially, frequencies of 2-mers, 3-mers and 4-mers were extracted from public databases of PCTs and lncRNAs of *Homo sapiens*, resulting in a dataset composed of two groups of RNA sequences, one for PCTs and the other for lncRNAs, and a large number of features. To reduce the number of features, S^2FS was applied to the dataset. Experimental results showed that relevant features were selected, keeping the predictive accuracy, with a lower processing cost than some existing feature selection techniques.

Keywords: Feature selection · Machine learning · lncRNAs · PCTs · Bioinformatics

1 Introduction

A considerable large part of eukaryotic genomes is composed of DNA portions that do not code for proteins. These molecules, known as non-coding RNAs (ncRNAs) [10], have important functions in the cell [8]. There are several types of

© Springer Nature Switzerland AG 2018
R. Alves (Ed.): BSB 2018, LNBI 11228, pp. 103–113, 2018.
https://doi.org/10.1007/978-3-030-01722-4_10

ncRNAs, classified according to their size and function within the cell, e.g., transfer RNAs (tRNAs), ribosomal RNAs (rRNAs), snoRNAs, microRNAs and many others [3]. These are known as small ncRNAs. Another class, called long noncoding RNAs (lncRNAs), is very heterogeneous, is longer than 200 nucleotides, and has a low protein coding capacity. Although their functional roles are still largely unknown [11], the interest in these molecules has increased in recent years. This occurred mainly because they regulate gene expression [17] and are associated with several diseases [13].

Currently, there are no widely accepted computational tools that address the problem of distinguishing lncRNAs from protein coding transcripts (PCTs) considering only their primary sequences. Therefore, this problem is a good candidate for the use of machine learning (ML) algorithms. In fact, some tools, using ML algorithms, have been developed to predict lncRNAs [5], e.g., lncRNApred [12], DeepLNC [19], lncRScan-SVM [5] or the method proposed by our group [18].

These tools use distinct machine learning (ML) algorithms and slightly different subset of features, since an optimal set of features, in different species, is currently not known.

The intuitive approach of combining all the features from different tools to build one single set of features, trying to obtain classification models with better predictive performance, does not hold. One reason is that the higher the ratio between the number of features and the number of instances, the higher the chances of overfitting, due to the curse of dimensionality, or Hughes phenomenon [6]. According to this phenomenon, after having reached the optimal number of features leading to good classifiers' performance, if the dimensionality is further increased, the more sparse the data becomes and, as a result of it, the model loses generalization ability. In fact, the number of training instances needed to induce predictive models with high generalization ability grows exponentially with the number of features.

To overcome this problem, there are two alternatives: increasing the number of training data exponentially, usually unfeasible, or reducing the number of features. In the latter, two approaches for dimensionality reduction can be used: feature selection, which chooses a subset of the best features to represent the original data; and feature aggregation, which combines and transforms the original features [2].

Since in this work it is important to keep biological information of lncRNAs and PCTs sequences, feature selection was chosen. An exhaustive selection algorithm choosing the best subset of features is usually unfeasible for high dimensional data. Thus, techniques have been developed to select a good set of features with a feasible computational cost [7]. The main feature selection techniques are based on one of the approaches - wrapper, filter, and embedded.

Besides, the techniques either rank the features according to their discrimination ability, or select a good subset of features that, together, lead to a higher discrimination ability. The second is a combinatorial optimization approach, leading to a high computational cost. This cost can be prohibitive if the number of features is large, as is the case of genome sequence datasets.

After reducing the number of features, a classification ML algorithm can be applied to the new dataset, inducing a predictive model about to discriminate instances from the different classes.

This work proposes an empirical assessment of a simple and fast feature selection technique, called S^2FS, to be used in ML algorithms, for distinguishing lncRNAs from PCTs. S^2FS uses a greedy heuristic to select features based on the classification performance of each individual feature, which is not correlated to any other already selected feature. The method is based on the assumption that a good subset contains features that are highly correlated with the response class, but are uncorrelated with each other [4].

This paper is organized as follows. First, in Sect. 2, we propose the S^2FS technique. Then, the results obtained using this technique are discussed in Sect. 3. Finally, in Sect. 4, we conclude and suggest future work.

2 The S^2FS Technique

In this section, we initially present the datasets used, then, we describe the S^2FS technique. Finally, we detail how the experiments were performed.

2.1 Datasets

The sequences used to build the datasets were downloaded from LNCipedia [20] and RefSeq [15]. The *Homo sapiens* dataset is composed of 111,145 protein coding transcripts (PCTs) from RefSeq, and 100,849 ncRNA transcript sequences from LNCipedia 4.0, both of the GRCh38 assembly.

Two datasets were initially constructed, the first with transcripts selected with an entropy-based clustering algorithm, while the second was produced with random samples from each class (PCTs and lncRNAs). Both subsets had the same number of transcripts for each class. These datasets were created to verify the accuracies of the models induced using different samples.

Since lncRNA is a very heterogeneous class, there might be transcripts not distinguishable from PCTs, when extracting features from their primary sequences. Therefore, the outlier transcripts may affect the decision boundary of the model. Some works, e.g., Tripathi et al. [19], have used sample selection to develop their classification models.

On the other hand, a model built only with transcripts that can be easily discriminated, obtained by sample selection methods, may lose the generalization capacity of classifying new transcripts. The previous work proposed by our group [18] used the random sample selection technique.

In the experiments, the sequences were divided into a training set, used to construct the model, and a test set, to evaluate the model predictive performance for new sequences.

Entropy-Based Clustering Selection. For the set of selected transcripts, an align-free approach, similar to Tripathi et al. [19], was used. Thus, four features were generated based on the Shannon Entropy (Eq. 1) of the occurrences of k-mers of sizes 2, 3 and 4 for each sequence. In this equation, p_i stands for the probability of occurrence of the k-mer i in the corresponding sequence.

$$H = -\sum_i p_i \log p_i \tag{1}$$

The sets of transcripts were built with pairs of sequences from distinct classes, farthest apart from each other, using these entropy features. The transcript selection method used the k-means algorithm [9] to group the transcripts in two clusters. The cluster mainly composed of lncRNAs was called *lncRNA-Cluster*, the other one *PCT-Cluster*. In each cluster, the class with the lowest number of transcripts, respectively PCTs and lncRNAs, was discarded. This is due to the fact that this method selects transcripts feature-wise similar to the other transcripts of the same class. Since the discarded transcripts are significantly different from their respective classes, they are not good samples for this method.

After creating and labeling the clusters, a transcript was chosen based on the largest Euclidean distance from the center of the opposite cluster (Fig. 1). After selecting the sequences of each dataset, the features used to select the sequences were discarded for the corresponding subset, in order to avoid introducing bias from this first step of selecting sequences in the process of feature selection.

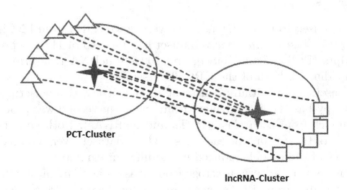

PCT-Cluster

lncRNA-Cluster

Fig. 1. The stars in the middle of the circles represent the cluster center, and the dotted lines from the stars to the circles' frontiers show the corresponding Euclidean distances.

Random Transcripts. After analyzing the results obtained with the datasets previously described, a more general set of samples was constructed. Therefore, a randomly generated dataset was created with the transcripts from RefSeq and LNCipedia 4.0.

In this dataset, even transcripts with 'N' (for not identified nucleotides) in their sequences were included in the set. These transcripts allowed us to verify how the feature selection techniques would deal with data noise.

2.2   Single Score Feature Selection

The S^2FS technique selects a subset of features based on their individual predictive performance in the classification of the training sequences. For such, each feature of the training sequences, i.e., each k-mer frequency with size 2, 3 or 4, was used to train a simple Gradient Boosting Classifier (GBC). The hyperparameters of this classifier were set by grid search, using 5-fold cross-validation. Figure 2 presents the pipeline of the ML algorithm, and details of S^2FS.

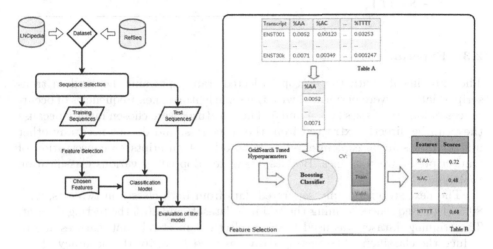

Fig. 2. Classification process with GBC, using one feature at a time. Table A shows the k-mers frequencies, for each transcript. Table B stores the obtained results, having been used as the choice criterion for the feature selection technique.

During the training process, an AUC score was obtained for experiments with each feature. Features with predictive performance not better than a random chance, i.e., AUC score lower than or equal to 0.5, were removed from the feature set. Next, the chosen features were classified based on a score, which can be either the AUC scores or a weighted Euclidean distance from $(0, 0)$ to (pc_hits, nc_hits), where pc_hits is the number of PCTs correctly classified by the feature, and nc_hits, the number of lncRNAs correctly classified. The classification accuracy together with the resulting features determine which score strategy is the best to be used. In the presented results, the AUC score was used for being faster, and also for achieving a feature set similar to the set generated by the other strategy (the weighted Euclidean distance score).

Once sorting the features by the scores, a greedy algorithm chooses the subset based on the AUC scores, avoiding features highly correlated with any other feature already selected for the set (Algorithm 1). Considering that all the features in this work are non-negative, the Pearson correlation was used to quickly check for linear correlation between the features.

Algorithm 1. Greedy selection with a correlation filter

Let $F[1 \ldots n]$ be a list of features f_i sorted by AUC_score
$S \leftarrow \{f_1\}$
for $i = 2$, step 1, until $i \leq n$ **do**
　$addFeature \leftarrow TRUE$
　for all $f_{s_j} \in S$ **do**
　　if $|\rho(f_i, f_{s_j})| > CorrelationThreshold$ **then**
　　　$addFeature \leftarrow FALSE$
　if $addFeature$ **then**
　　$S \leftarrow S + \{f_i\}$

2.3 Experiments

The experiments with the entropy selected transcripts and the random transcripts datasets were performed with the same features, i.e., frequencies of occurrences of k-mers of sizes 2, 3, 4 and 5. These features were chosen mainly because they can be directly extracted from the sequences, and do not need any other information source, e.g., interaction with other transcripts or conservation of transcripts between species. Besides, they are adopted by various classification tools [5].

The next step randomly separated data from both classes in two sets, with 80% of the sequences forming the training dataset, and 20% the testing dataset. The training dataset was used both to select the most relevant features and to induce the classifiers. The testing data was used to verify the accuracy of the classification model. Besides, to verify the importance of each feature chosen by the selection techniques, the Random Forest algorithm [1] was used to induce the classifiers.

The experiments used sets from 5 to 30 features, with a step of 5. In order to avoid introducing errors in the measurements, all the experiments were run on the same machine. The experiments were repeated 40 times, with different seeds to split data between training and testing sets. The predictive results obtained for the selected features were the average of all these executions. The results were validated using statistical tests with a confidence interval of 90%.

The predictive performance of the classifiers using features selected with S^2FS was compared to those with random selection and also to two other feature selection techniques: an univariate selection technique; and a Sequential Forward Selection (SFS) technique [16]. The univariate feature selection is similar to S^2FS in the sense that it investigates each feature individually, using a statistical test to determine the relationship strength between the feature and the response variable. In this work, the χ^2 test function was used as the scoring function to select features. On the other hand, SFS is a more strict technique. In the first step, similarly to S^2FS, it tests every feature with a specified classification function. However, in the next steps, it tests the new features to be added alongside with the already selected features on the classification function. This is potentially a very resource intensive task, depending on the classification function used. The more precise the classification function, the more time is needed to find the features. In this work, the k-Nearest Neighbors function was used, since it provided the best balance between accuracy and speed of the feature selection process.

The final experiment with S^2FS used a larger set of random selected transcripts. The proposed technique selected 300 from a set of 1,642 features, composed of: the frequencies of k-mers of sizes 2, 3, 4 and 5; features of entropy - H2, H3, H4 and H5; the longest ORF size; the GC content; the ORF coverage; the ORF start position in the transcript; and the coefficient of variation of position of 4-mers inside the longest ORF (CV_positions).

Then, these selected features were used to train two classification models with two different algorithms: a Gradient Boosting Classifier (GBC) and a Multilayer Perceptron (MLP) [14]. The GBC was set with a learning rate of 0.1, 100 estimators and a max depth of 6 features. The MLP was set with two layers with 512 neurons, a RELU activation function, dropout regularization of 0.65 between the layers, and used the Adam optimization function combined with an early stop strategy for training the model.

3   Results

The first experiments were performed with a set of all the k-mers, $k = 2, 3, 4$ and 5, from the selected sets of transcripts. The objective was to assess if S^2FS would find a good subset of features from a set of transcripts relatively easy to distinguish. The results are presented in Table 1, Fig. 3. Since all the feature selection techniques used the same classifier, the results are labeled by the corresponding feature selection technique.

Table 1. Accuracies obtained with four different feature selection methods, taking as input the entropy based selected transcripts of PCTs and lncRNAs of *H. sapiens*.

Features	Random	Univariate	SFS	S^2FS
5	0.97427	0.94009	0.99422	0.99230
10	0.99109	0.97913	0.99720	0.99602
15	0.99634	0.98725	0.99795	0.99733
20	0.99742	0.99049	0.99844	0.99790
25	0.99817	0.99177	0.99858	0.99807
30	0.99847	0.99387	0.99886	0.99856

According to Fig. 3, both SFS and S^2FS accurately predicted more than 90% of the testing instances using only 15 features. Their accuracies slightly increased as more features were added. In general, they were much better than the predictive performance of the two other techniques. On the other hand, the univariate technique did not perform well with this subset, having presented accuracies

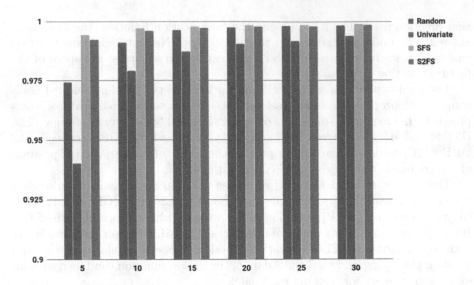

Fig. 3. Predictive performance of the four feature selection techniques, for the same selected set of transcripts of *H. sapiens*.

worse than the random choice technique in all the tests. This indicates that the feature selection technique strongly affects the model predictive accuracy.

Thus, trancripts' selection seemed to have a strong effect in producing several good features. Even the random approach, when selecting 25 features or more, could select a set that enabled the classifier to obtain an accuracy of more than 90%. However, there was no evidence that a model trained with features extracted from these selected transcripts would be able to generalize transcripts' classification. To verify this hypothesis, the set with the random selected transcripts was tested. The obtained accuracies were significantly lower, as shown in Fig. 4 and comparing the results from Tables 1 and 2.

Table 2. Accuracies obtained with four different feature selection methods, taking as input random selected transcripts of PCTs and lncRNAs of *H. sapiens*.

Features	Random	Univariate	SFS	S^2FS
5	0.76168	0.69068	0.69581	0.82266
10	0.83247	0.83073	0.73423	0.85812
15	0.84766	0.86219	0.74640	0.87114
20	0.86078	0.87513	0.75002	0.87895
25	0.86607	0.87975	0.75581	0.88231
30	0.86898	0.88786	0.75630	0.88420

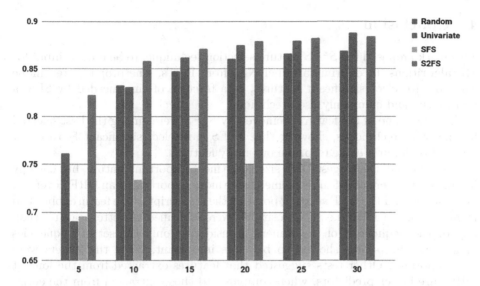

Fig. 4. Performance of the four feature selection techniques, for the random set of transcripts of *H. sapiens*.

On the set with random transcripts, the univariate selection technique performed better, and was able to select the best set of 30 features, when compared to the other techniques. On the other hand, SFS performed poorly. When analyzing the features selected by SFS, several features containing N were selected. This means that SFS, with the chosen base classifier, did not perform well in the presence of noise in the data. S^2FS found the best set of features for sets with 10, 15, 20 and 25 features, and did not select any of the noise features of the random set.

An interesting aspect of the S^2FS is that when it chooses features from a larger and heterogeneous set of frequency features (2-, 3-, 4- and 5-mers), it chooses mainly the frequencies of 5-mers and 4-mers. These are also the most significant features chosen by the models trained with Random Forests.

Although the execution time was not the objective of this work, S^2FS performed much faster than SFS. While SFS needs 14 h to select a set of 30, from 336 features, S^2FS selected 30 features in less than 1 h. This reduction in processing time is a good contribution, especially considering projects with very large input features.

The last experiment was performed using S^2FS to select 300 features, from a set of 1,624, for the same dataset: 16 2-mers; 64 3-mers; 256 4-mers; 256 CV_positions of 4-mers; 1024 5-mers; H2, H3, H4 and H5; longest ORF_size, GC content, ORF_coverage and ORF_Star_pos. It reached an accuracy of 0.9752 using MLP and GBC. These algorithms use different methods to build classification models, but the selected features provided good accuracy with both methods.

4   Conclusion

This work proposed the S^2FS feature selection technique, to be used as input for ML algorithms, to discriminate lncRNAs from PCTs. The proposed technique was able to select significant features, in a fraction of time needed by SFS, a more strict and commonly used technique.

Besides, none of the features removed by S^2FS were among the best selected by the other techniques, showing that S^2FS can select significant features, or be used on its first stage to prune out noisy features.

As expected, the longest ORF size is the most important feature, but entropy features for 5-mers and 4-mers seemed to be more important than ORF_coverage, GC_content and the ORF starting point on the transcripts. The features obtained from the longest ORF, i.e., CV_positions, were identified as better features than any of the frequencies of the k-mers. Considering only the set of frequencies of k-mers, the 5-mers showed to be more important, while the 2-mers were less important. Other tests suggested that features extracted from the longest ORFs are better predictors, when compared to those extracted from the entire transcript.

Future works include testing S^2FS with a more heterogeneous set of features of lncRNAs and PCTs. Also, it is possible to explore processors' parallelism, since each feature is evaluated separately, with a distinct thread (or process), which could accelerate the initial process of feature selection. Other classifiers could also be tested in the first step of S^2FS, to compare the resulting set of features. Since S^2FS was able to fast remove noisy features in the first step, it could be used in an Ensemble of feature selectors combining different techniques with different ML classifiers, to find better feature subsets on a reasonable amount of time.

References

1. Breiman, L.: Random forests. Mach. Learn. **45**(1), 5–32 (2001)
2. Cai, J., Luo, J., Wang, S., Yang, S.: Feature selection in machine learning: a new perspective. Neurocomputing **300**(26), 70–79 (2018)
3. Esteller, M.: Non-coding RNAs in human disease. Nat. Rev. Genet. **12**(12), 861 (2011)
4. Hall, M.A.: Correlation-based feature selection for machine learning. Ph.D. thesis, University of Waikato Hamilton, April 1999
5. Han, S., Liang, Y., Li, Y., Du, W.: Long noncoding RNA identification: comparing machine learning based tools for long noncoding transcripts discrimination. BioMed Res. Int. 2016 (2016)
6. Hughes, G.: On the mean accuracy of statistical pattern recognizers. IEEE Trans. Inf. Theory **14**(1), 55–63 (1968)
7. Jain, A., Zongker, D.: Feature selection: evaluation, application, and small sample performance. IEEE Trans. Pattern Anal. Mach. Intell. **19**(2), 153–158 (1997)
8. Kaikkonen, M.U., Lam, M.T., Glass, C.K.: Non-coding RNAs as regulators of gene expression and epigenetics. Cardiovas. Res. **90**(3), 430–440 (2011)

9. Lloyd, S.: Least squares quantization in PCM. IEEE Trans. Inf. Theor. **28**(2), 129–137 (2006). https://doi.org/10.1109/TIT.1982.1056489

10. Mattick, J.S.: Non-coding RNAs: the architects of eukaryotic complexity. EMBO Rep. **2**(11), 986–991 (2001)

11. Mattick, J.S., Rinn, J.L.: Discovery and annotation of long noncoding RNAs. Nat. Struct. Mol. Biol. **22**(1), 5 (2015)

12. Pian, C., et al.: LncRNApred: classification of long non-coding RNAs and protein-coding transcripts by the ensemble algorithm with a new hybrid feature. PloS One **11**(5), e0154567 (2016)

13. Ponting, C.P., Olive, P.L., Reik, W.: Evolution and functions of long noncoding RNAs. Cell Volume **136**(4), 629–641 (2009)

14. Popescu, M.C., Balas, V.E., Perescu-Popescu, L., Mastorakis, N.: Multilayer perceptron and neural networks. WSEAS Trans. Circ. Syst. **8**(7), 579–588 (2009)

15. Pruitt, K.D., Tatusova, T., Maglott, D.R.: NCBI reference sequences (RefSeq): a curated non-redundant sequence database of genomes, transcripts and proteins. Nucleic Acids Res. **35**(Suppl. 1), D61–D65 (2007)

16. Pudil, P., Novovičová, J., Kittler, J.: Floating search methods in feature selection. Pattern Recogn. Lett. **15**(11), 1119–1125 (1994)

17. Rinn, J.L., Chang, H.Y.: Genome regulation by long noncoding RNAs. Ann. Rev. Biochem. **81**, 145–166 (2012)

18. Schneider, H.W., Raiol, T., Brigido, M.M., Walter, M.E.M., Stadler, P.F.: A support vector machine based method to distinguish long non-coding RNAs from protein coding transcripts. BMC Genomics **18**(1), 804 (2017)

19. Tripathi, R., Patel, S., Kumari, V., Chakraborty, P., Varadwaj, P.K.: DeepLNC, a long non-coding RNA prediction tool using deep neural network. Netw. Model. Anal. Health Inform. Bioinform. **5**(1), 1–14 (2016)

20. Volders, P.J., et al.: LNCipedia: a database for annotated human lncRNA transcript sequences and structures. Nucleic Acids Res. **41**(D1), D246–D251 (2013)

A Genetic Algorithm for Character State Live Phylogeny

Rafael L. Fernandes[1]([✉]), Rogério Güths[2], Guilherme P. Telles[3],
Nalvo F. Almeida[2], and Maria Emília M. T. Walter[1]

[1] Departamento de Ciência da Computação, Universidade de Brasília,
Brasília, Brazil
leafarlins@gmail.com, mariaemilia@unb.br
[2] Faculdade de Computação, Universidade Federal de Mato Grosso do Sul,
Campo Grande, Brazil
r.guths@ufms.br, nalvo@facom.ufms.br
[3] Instituto de Computação, Universidade Estadual de Campinas, Campinas, Brazil
gpt@ic.unicamp.br

Abstract. Character state live phylogeny generalizes character state phylogeny in the sense that they relate taxonomic units based on their similarities over a set of characters, but allowing live ancestors. An approach for character state live phylogeny reconstruction is called parsimony, where one tries to minimize the total number of character state changes along the edges of the tree. The problem of finding a tree that minimizes this number is known as large live parsimony problem. When the tree topology is also given as input, the problem is known as small live parsimony problem. We propose a genetic algorithm to solve the large live problem, which uses extended versions of the algorithms of Fitch and Sankoff to solve the small live problem, both devised in this work. Besides, we performed two experiments. In the first one, a multiple alignment of H1N1 and H3N2 viruses from different countries, taken as input, allowed to obtain interesting live phylogenies, representing alternative evolutionary hypothesis. The second experiment took as input a multiple alignment of the HIV virus *env* gene, from one patient, read in different dates through 12 years. The generated live phylogenies were similar to the ones generated by PAUP, where dates close to each other were grouped into clusters, but suggesting new evolutionary stories.

Keywords: Live phylogeny · Genetic algorithms · Parsimony
Sankoff algorithm · Fitch algorithm

1 Introduction

Phylogeny reconstruction aims at finding evolutionary relations among objects, which are shown by a tree elucidating how these objects are related to each other through common ancestors. In these trees, internal nodes represent hypothetical

© Springer Nature Switzerland AG 2018
R. Alves (Ed.): BSB 2018, LNBI 11228, pp. 114–123, 2018.
https://doi.org/10.1007/978-3-030-01722-4_11

ancestors, while the objects are shown in the leaves. Several methods have been proposed for reconstructing phylogenies [3, 10].

Input data for phylogeny reconstruction are of two types: (i) discrete characters with a finite number of states, formalized in a character state matrix $M_{objects \times characters}$, where $M[i, j]$ holds the state of character j of object i; and (ii) comparative numerical data, represented in a distance matrix $D_{objects \times objects}$, where $D[i, j]$ holds the distance between objects i and j.

Telles et al. [13] generalized the concept of phylogeny for both types of input data by allowing the existence of taxonomic objects as ancestors. This concept, called Live Phylogeny, suits the case of fast-evolving species, e.g., virus, and also the construction of phylogenies for non-biological objects, e.g., documents, images, and database records.

Two problems related to live phylogeny based on character states are known [5], and are enunciated next. The *small live parsimony problem* is: given a set of objects, a character state matrix and a phylogeny allowing live ancestors as input, find a labeling of the internal nodes with a minimum number of changes of states along the given topology. The *large live parsimony problem* is: given a set of objects and a character state matrix as input, find a phylogeny and a labeling of internal nodes, live ancestors allowed, which minimizes the total number of state changes.

As their classical counterparts, the small live parsimony problem may be solved in polynomial time, and the large live parsimony problem is an NP-hard problem. In this article, we propose the usage of genetic algorithms to deal with practical instances of the large live parsimony problem, thus allowing the construction of alternative hypothesis for the relationship among the input objects.

The text is organized as follows. In Sect. 2, we propose a genetic algorithm for the large live parsimony problem, using algorithms to solve the small live parsimony problem. For this last one, we extended the algorithms of Fitch and Sankoff to deal with live phylogenies. In Sect. 3, we discuss the results obtained in two case studies. Finally, in Sect. 4, we conclude and suggest future work.

2 A Genetic Algorithm for Live Phylogeny

Genetic algorithms [8] (GAs) are created as an analogy to biology, with three basic ideas: (i) the maintenance of a population of possible solutions; (ii) the selection of individuals of the population by a fitness function; and (iii) the application of operators which generate mutations and recombinations throughout several generations.

GAs are a metaheuristic to solve hard problems, which would require an exhaustive search in a large solutions space to find an optimal one. GAs were applied to phylogeny reconstruction with different choices of population generation, fitness function and operators [6, 7, 14].

Our genetic algorithm for large live parsimony phylogeny construction is called GA-LP. Figure 1 outlines the steps of GA-LP. Next sections discuss some steps in more detail. The input is a character state matrix.

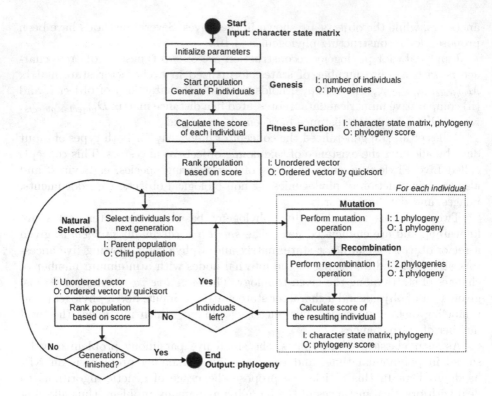

Fig. 1. Outline of GA-LP, a genetic algorithm for live phylogeny. Here, I and O refer to the *Input* and *Output*, respectively, of each module.

Some parameters can be set for GA-LP: the population size; the fraction of best individuals in the population to be preserved for the next generation; the probability decay for selecting other individuals for the next generation; the probabilities of mutation and recombination operators; the fraction of live internal nodes in the initial population; the range of live ancestors in the final tree; the fitness function; and the stop criterion.

The initial population is built by randomly generating trees with a maximum number of live ancestors. Either Live Fitch algorithm or Live Sankoff algorithm (described next) may be used to label the internal nodes and to score live phylogenies in the population. At each iteration, selected parents create offspring for the next generation. GA-LP uses a combination of elitism and randomization for such selection.

Mutations and recombination are based on previous GAs for phylogeny [6, 14], but modified for live phylogenies. A new mutation operator is introduced to generate live ancestors or leaves, as detailed below.

GA-LP stops when either a maximum number of generation is reached, or when the best score of the population reaches a value less than a threshold, or when the best score of the population does not change in a fixed number of generations. At the end, GA-LP returns the best phylogeny in the last generation.

2.1   Fitness Functions

The small phylogeny problem can be solved by the algorithms of Fitch [4] and Sankoff [9]. Both algorithms label the internal nodes of a tree according to the states of its leaves, which are given as input, and both algorithms have a bottom-up and a top-down phase. We refer the reader to the bibliography for further details on these algorithms. Below we show how the bottom-up phases of Fitch and Sankoff algorithms may be modified to allow live nodes.

Algorithm 1 shows Fitch's algorithm adapted to allow live ancestors. The modification was the inclusion of live ancestors, for which the score can be raised by 1 or 2, depending on the two descendants. It means that, if both descendants do not have the same state of node i, it is necessary two evolutionary changes to use the tree as the evolutionary history of the objects.

Algorithm 1. Live Fitch bottom-up phase

Input: $M_{n \times m}$ matrix and phylogeny T (n objects, m characteristics)
Output: $finalscore$ integer

```
 1:  finalscore = 0
 2:  for each column m of the input matrix do
 3:      for each node labeled by s_k do
 4:          R_k = {s_k}, R the set of possible states
 5:      end for
 6:      score = 0
 7:      for each node i from leaves to root in post-order do
 8:          if node i is a hypothetical ancestor with children u and v then
 9:              if R_u ∩ R_v ≠ ∅ then
10:                  R_i = R_u ∩ R_v
11:              else
12:                  R_i = R_u ∪ R_v
13:                  score = score + 1
14:              end if
15:          else
16:              ▷ node i is a live ancestor with children u and v
17:              R_i = {s_i}
18:              if R_i ∩ R_u = ∅ then
19:                  score = score + 1
20:              end if
21:              if R_i ∩ R_v = ∅ then
22:                  score = score + 1
23:              end if
24:          end if
25:      end for
26:      finalscore = finalscore + score
27:  end for
```

For the Sankoff's algorithm, it is necessary a matrix $Cost_{p \times p}$ with the costs of changing each pair of states, for all the p states, where $Cost(a, b)$ denotes the cost of changing state a to b. A cost c is associated to each node, where $c(i, a)$ is the cost of labeling node i with label a, which is one of the states. The state of node i is denoted by s_i.

Similarly to Fitch, Algorithm 2 shows Sankoff's algorithm adapted to allow live ancestors. In this case, the modification was the inclusion of live ancestors as internal nodes. The algorithm calculates the cost of labeling the node i with s_i already defined, keeping the initial value of ∞ in the other labels.

Algorithm 2. Live Sankoff bottom-up phase

Input: $M_{n \times m}$ matrix, $Cost_{p \times p}$ matrix and phylogeny T (n objects, m characteristics, p states)
Output: integer $finalscore$

```
1:  finalscore = 0
2:  for each column m of the input matrix do
3:      for each node i with defined label s_i = a do
4:          c(i, l) = ∞ for each l ≠ s_i
5:          c(i, a) = 0
6:      end for
7:      for each node i, from leaves to root in post-order, do
8:          if node i is a hypothetical ancestor then
9:              for each label l_k, 1 ≤ k ≤ p do
10:                 c(i, l_k) = ∑_{v is child of i}(Cost(l_k, b_v) + c(v, b_v))
11:                 such that b_v = min{c(v, r_j) + Cost(l_k, r_j)}, 1 ≤ j ≤ p
12:             else
13:                 ▷ node i is a live ancestor
14:                 for l such that s_i = l do
15:                     c(i, l) = ∑_{v is child of i}(Cost(l, b_v) + c(v, b_v))
16:                     such that b_v = min{c(v, r_j) + Cost(l, r_j)}, 1 ≤ j ≤ p
17:             end if
18:         end for
19:         finalscore = finalscore + minimum value of c(root, l)
20: end for
```

2.2 Natural Selection, Mutations and Recombination

GA-LP uses a natural selection based on ranking. After sorting by score the population, with length N, the first individual is protected from mutation and the elitist strategy[1] copies the best E individuals to the next generation. Then it selects $N - E$ individuals from the remaining list through a probability function that depends on the ranking. This probability function can be adjusted, behaving as a homogeneous function or favouring the selection of individuals with higher score at various degrees.

According to the probabilities set in the parameter initialization, GA-LP randomly decides if each individual is mutated or recombined. There are three mutation operators on a single tree: *Mutation 1* switches two disjoint subtrees selected at random; *Mutation 2* switches the label of two nodes selected at random; and *Mutation 3* moves live nodes or leaves either up or down, allowing to create a live node when a leaf is moved up, or to remove a live node when it is moved down and becomes a leaf, thus leading to different configurations of a set of live nodes. This last mutation was designed for live phylogeny, with possible situations illustrated in Fig. 2. *Recombination 1* interchanges parts of two individuals, one from the parent and the other from the offspring generation randomly chosen, and is based on a GAML [6] operation.

3 Results

We performed experiments both to calibrate the parameters and to evaluate the performance of GA-LP on real data. They were executed on a computer with an Intel i5 processor and 4 GB RAM.

---

[1] The elitist strategy creates a new population transferring the best organisms, not modified, from the current generation to the next.

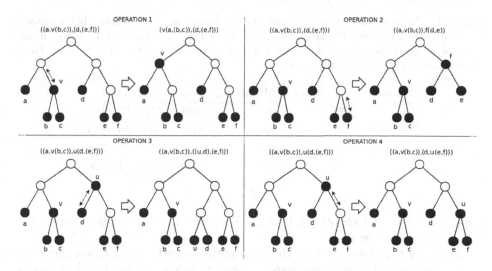

Fig. 2. Examples for Mutation 3, specially designed for live phylogeny.

3.1 Parameter Calibration

To calibrate the parameters, we used a matrix $M_{34 \times 306}$, with 34 different organisms and 306 characteristics, which are the columns of a multiple sequence alignment of proteins (20 states). We executed the algorithm 10 times and obtained the average and the standard deviation of the output values. Initially, we evaluated the behaviour of each mutation. We noted that Mutation 1 causes a steeper reduction of the best score. Also, Recombination 1 decreases the average score of the population faster.

We analyzed the output of different combinations of varying rates of the three mutations, the recombination rate, and also the FV parameter, which defines the ratio of live ancestors in the initial population. The stop criterion was a score lower than 130. This value was defined experimentally, performing a large number of executions of the algorithm. Once the minor value reached for the input, after a huge times of executions, was the score of 128, we have arbitrarily chosen the score of 130. The best results are shown in Table 1. Experiments were executed with all combinations of mutation rates in $\{0, 25, 50, 75\}$, fixing the other parameters.

The results showed that Mutation 1 should be as high as possible to obtain the best combination. We fixed the rates of Mutation 1, 2 and 3 for the next experiment to 70, 15 and 15, respectively. In the sequel, we found the best value of the recombination rate, changing its values. As soon as we increased the recombination rate above 50, the number of generations also increased. The smallest values were reached with rates equal to 40 or 50.

Best results were achieved when the elite parameter E was set at least equal to 10% of the population size. For lower values of E, the number of generations became much higher. Thus, for some experiments, we fixed E to 10% of the

Table 1. The first three columns show the combinations of mutations that achieved the best results. Column *Generations* shows the average and standard deviation of the number of generations needed to reach a score lower than 130, while column *Time* shows the total time of 10 executions.

Mutation 1	Mutation 2	Mutation 3	Generations	Time
75	25	0	1945 ± 1321	07:42:06
75	0	25	2524 ± 1660	10:08:44
50	25	25	2566 ± 1323	10:07:48
50	25	25	2920 ± 1446	11:11:39

population size, and for others, to 20%. We note that this is a parameter that can be chosen by the user.

We also set the parameter for the range of the desired number of live ancestors in the phylogeny. The other software parameters were fixed, varying the range of desired live ancestors, from 1 to 3, from 4 to 6, and so on, up to the range 16 to 18. The stop criterion was set with a parameter that interrupts the execution if 500 generations were performed without changing the best score of the population in this experiment. In these experiments, the final score increases as more live ancestors are required. Thus, the parameter forcing the existing of live ancestors is important.

3.2   Two Case Studies

H1N1 and H3N2 Viruses. In this case study, the input was a matrix $M_{75 \times 759}$, with 75 organisms and 759 characteristics, of a multiple sequence alignment of the RNA polymerase PB2 from H1N1 and H3N2 viruses (4 states), collected in 2016 at four countries (USA, Russia, China, South Korea), and downloaded from the NCBI influenza virus sequence database (ftp://ftp.ncbi.nih.gov/genomes/INFLUENZA/).

We generated two phylogenies with GA-LP using the parameters described previously, both including and not including live ancestors, and we compared these results to a phylogeny generated with the Proml program of Phylip [1].

Figure 3 shows part of the tree without live ancestors, where each node is represented by a number, country, region of origin and type of virus.

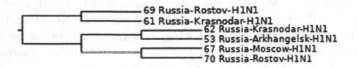

Fig. 3. Part of the phylogeny, showing objects from Russia in a cluster, obtained by the GA-LP with no live ancestors allowed.

Figure 4 shows part of the tree with live ancestors. We note that the clusters generated by GA-LP were similar to those generated by Phylip, in the sense that viruses found in geographically close regions were usually grouped in clusters.

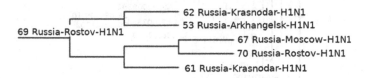

Fig. 4. Part of the live phylogenetic tree, showing the Russia cluster, obtained by the GA-LP allowing live ancestors.

We noted that the final score increases with a higher number of live ancestors, linearly for $M_{34 \times 306}$, and non-linearly for $M_{75 \times 759}$. It means that phylogenies with larger number of live ancestors have usually a worse parsimony, when compared with phylogenies with few or no live ancestors.

The *env* Gene of the HIV Virus. The other case study used as input two matrices, $M_{52 \times 765}$ and $M_{136 \times 680}$, with data correspondent to multiple sequence alignments from the C2-V5 sector of *env* gene of the HIV virus, all from the same patient, read in different dates (timepoints) through 12 years, cited by Shankarappa [11], and available at the HIV Databases web site [2]. The Sankoff's algorithm was used as the fitness function. A variable cost matrix for the gene characters of the reads was used.

We selected three output phylogenies to analyze, with 0, 18 and 54 live ancestors. These phylogenies were compared, as well as a tree generated by PAUP [12]. The phylogenies showed similar clusters, with objects closed to each other if they were read in close dates (timepoints). Figure 5 shows parts of the tree, with clusters generated from objects of the timepoints 11, 12, 13 and 15.

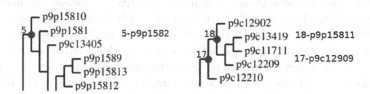

Fig. 5. Parts of the 18 live ancestors phylogeny, with objects representing reads from timepoints 11 to 15 in a same cluster.

Figure 6 shows part of a tree with 54 live ancestors, where the number of live ancestors (54) is higher than the number of hypothetical ancestors (13). We can see a large cluster with objects corresponding mostly to the timepoints 00 and 01.

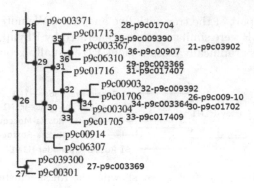

Fig. 6. Parts of the 54 live ancestors phylogeny, with objects representing reads from timepoints 00, 01 and 03.

In general, the clusters generated with GA-LP were similar to those generated with PAUP. It is noteworthy that the live phylogeny shows us a different evolutionary history of the input objects. These experiments indicated, mainly due to the method used to calculate the score, that a proper definition of the range of live ancestors is essential to obtain good live phylogenies.

4 Conclusion

In this work, we presented a genetic algorithm to reconstruct a live phylogeny, called GA-LP, taking as input character state matrices. Also, we proposed and implemented a modified version of Fitch's and Sankoff's algorithms to solve the small live parsimony problem. With the implementation, after having calibrated the parameters, we performed two case studies. The first one took as input a character matrix obtained from a multiple sequence alignment of H1N1 and H3N2 from different countries. GA-LP constructed a tree that clusterized viruses from similar geographic regions. In general, this live phylogeny preserved clusters when compared to two other trees, produced by GA-LP generating a tree without live ancestors and by Phylip. The other case study was developed taken as input a character matrix constructed from a multiple sequence alignment of the HIV *env* region, read from one patient, through different timepoints. Output phylogenies were obtained with different numbers of live ancestors, with similar clusters, regarding to the objects' timepoints. Also, the phylogeny generated with PAUP showed clusters grouping objects with close timepoints, similar to those generated with GA-LP.

We intend to improve the parameters' calibration, exhaustively and using different datasets, taking care to not introduce overfitting. Besides, more experiments have to be performed, to further explore the advantages and the limitations of the approach, specially regarding the distance to the optimal tree, which may be found by branch-and-bound.

References

1. Baum, B.R.: Phylip: phylogeny inference package (version 3.2). Q. Rev. Biol. **64**, 539–541 (1989)
2. Data-HIV: HIV sequence database - special interest alignment set 1. https://www. hiv.lanl.gov/content/sequence/HIV/SI_alignments/set1.html. Acessed Apr 2018
3. Felsenstein, J.: Inferring Phylogenies. Palgrave Macmillan, Basingstoke (2004)
4. Fitch, W.M.: Toward defining the course of evolution: minimum change for a specified tree topology. Syst. Zool. **20**, 406–416 (1971)
5. Güths, R., Telles, G.P., Walter, M.E.M.T., Almeida, N.F.: A branch and bound for the large live parsimony problem. In: Proceedings of 10th International Joint Conference on Biomedical Engineering Systems and Technologies. BIOSTEC 2017, vol. 3, pp. 184–189 (2017)
6. Lewis, P.O.: A genetic algorithm for maximum-likelihood phylogeny inference using nucleotide sequence data. Mol. Biol. Evol. **15**(3), 277–83 (1998)
7. Matsuda, H.: Protein phylogenetic inference using maximum likelihood with a genetic algorithm. In: Proceedings of Pacific Symposium on Biocomputing, pp. 512–523 (1996)
8. Mitchell, M.: Introduction to Genetic Algorithms. The MIT Press, Cambridge (1999)
9. Sankoff, D.: Minimal mutation trees of sequences. SIAM J. Appl. Math. **28**, 35–42 (1975)
10. Setubal, J., Meidanis, J.: Introduction to Computational Molecular Biology. PWS Publishing Company, Boston (1997)
11. Shankarappa, R., Margolick, J.B., Gange, S.J., et al.: Consistent viral evolutionary changes associated with the progression of human immunodeficiency virus type 1 infection. J. Virol. **73**(12), 10489–10502 (1999)
12. Swofford, D.L.: Phylogenetic Analysis Using Parsimony (and Other Methods). Sinauer Associates, Sunderland (2002)
13. Telles, G.P., Almeida, N.F., Minghim, R., Walter, M.E.M.T.: Live phylogeny. J. Comput. Biol. **20**(1), 30–37 (2013)
14. Zwickl, D.J.: Genetic Algorithm Approaches for the Phylogenetic Analysis of Large Biological Sequence Datasets Under the Maximum Likelihood Criterion. Ph.D. thesis, The University of Texas (2006)

A Workflow for Predicting MicroRNAs Targets via Accessibility in Flavivirus Genomes

Andressa Valadares[1]([✉]), Maria Emília Walter[1], and Tainá Raiol[2]

[1] Department of Computer Science, Institute of Exact Science, University of Brasília, Brasília, Brazil
andressarodrial@gmail.com
[2] Fiocruz Brasília, Oswaldo Cruz Foundation, Brasília, Brazil

Abstract. Flavivirus infections are a serious public health issue in Brazil, particularly in recent years due to the large number and severity of cases of Zika and Dengue virus infections and, more recently, outbreaks of Yellow Fever virus infections. Therefore, understanding the effects of genetic variations at functional and structural levels and developing new tools are necessary for supporting arboviral surveillance and control efforts of these viruses. In this context, we developed a workflow to predict potential microRNA targets in Flavivirus genomes. The workflow implementation comprised the integration of Perl scripts, tools from ViennaRNA package, and miRanda software to search for potential microRNAs that potentially interact with non-coding regions of Flavivirus genomes. As a case study, genome sequences of Dengue virus serotypes were used. We could observe structural differences among the serotype sequences and miRNA target binding sites exclusively identified for each serotype, which may be useful for the development of diagnostic methods.

Keywords: Flavivirus · Dengue · microRNA
RNA secondary structure · Bioinformatics workflow

1 Background

The flaviviruses are the most common arthropod-borne viruses worldwide, which includes viruses transmitted by Dengue virus (DENV), West Nile virus (WNV), Yellow fever virus (YFV), Japanese encephalitis virus (JEV) [10] and Zika virus (ZIKV) [6]. These infections commonly cause febrile illness or syndromes such as encephalitis and hemorrhagic fever [24]. Despite the huge public health impact, there is no specific antiviral therapy available for treating any of the flaviviruses infections [7].

Dengue fever is an international public emergency due to its rapid spread and serious consequences [20]. It is estimated that annually 390 million infections occur worldwide, being more prevalent in developing countries [3]. The infection

R. Alves (Ed.): BSB 2018, LNBI 11228, pp. 124–132, 2018.
https://doi.org/10.1007/978-3-030-01722-4_12

is characterized by fever, rash, and in the more severe forms, hemorrhagic fever and shock syndrome [21]. This virus is transmissible by mosquitoes and presents four different serotypes (DENV-1, DENV-2, DENV-3, and DENV-4) [14].

In recent years, extensive efforts have been made into understanding the mechanisms behind flavivirus spread, replication, and pathogenesis [26]. For instance, it has been reported that microRNAs play an important role in viruses replication due to its ability to control viral and host gene expression [23].

MicroRNAs are small regulatory RNAs (~23nt) that play an essential part in gene regulation by binding to target mRNAs and suppressing their translation [2]. The accessibility of the target binding site is an important factor for determining the microRNA repression efficacy [12]. It is required at least four consecutive unpaired bases on target binding site to a successful microRNA-mRNA interaction [15]. Therefore, predicting conserved secondary structures is crucial to find more effective targets.

Despite the target site accessibility importance, only a few tools assume the role of the target secondary structure. For example, the PITA software that considers mRNA binding site structure affecting target recognition by thermodynamically promoting or disfavoring the interaction [12]. Although, this program has low efficiency compared to other algorithms [18]. Another limitation of the available prediction tools is that most of them are only web applications, performing the analysis of only few sequences at the same time and requires manual input.

However advances have been reached in recent years, the interaction between host microRNAs and flavivirus genomic RNA still requires further analysis [26]. Therefore, in this work, we present a workflow to predict microRNA targets by site accessibility for non-coding regions of flaviviruses. The workflow integrates available tools for secondary RNA and microRNA target predictions, combining the results, predicting accessibility, and suggesting potential microRNA targets.

2   Workflow for Prediction of MicroRNAs Targets via Accessibility

With the proposed workflow, secondary structures and microRNA targets are predicted within non-coding regions of flavivirus using full-length RNA genomic sequences (Fig. 1).

First, only non-coding regions of virus genomes are selected using developed Perl scripts. Full-length RNA sequences are grouped by non-coding region sizes (5′ and 3′ untranslated regions, UTR), then the most common length of each region was chosen for downstream analysis. Next, the 3′UTR sequences were split into three sub-regions (R1, R2, and R3) according to known domains of flavivirus genomes within this region [4]. After selection of non-coding regions, all redundancy is removed using the PRINSEQ software [22], followed by multiple sequence alignments using the MUSCLE software [8].

RNAalifold [17] from ViennaRNA Package was used to predict a consensus secondary structure of non-coding regions of flaviviruses, using default parameters. Next, structure graph representations were generated using VARNA [5] and

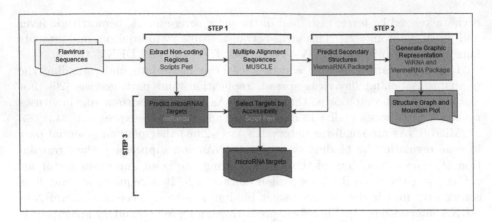

Fig. 1. Workflow to compare predicted secondary structures and to predict microRNA targets within genomic RNA sequences of Flavivirus via accessibility

mountain plots using cmount.pl. Finally, secondary structure predictions were performed for each sequence using refold.pl. The program RNAalifold and the scripts refold.pl and cmount.pl belongs to ViennaRNA Package [16]. The approach using tools from the ViennaRNA Package to predict secondary RNA structures of Flavivirus was partially based on a previous study [25]. In the present work, we improved the data analysis with additional and updated softwares, as well as employing different visualization tools of the predicted structures.

The microRNAs of primates extracted from miRBase [13] and the processed flavivirus sequences are inputs to the miRanda software [9]. Last, a Perl script was developed to select only accessible targets comparing the results from miRanda to the previously predicted structures. We consider accessible microR-NAs targets those with microRNA-mRNA complementary matches of at least four contiguous unpaired nucleotides within binding sites [15].

3   Case Study

The validation of the proposed workflow has been carried out by using Dengue virus sequences as input. Each step of the workflow was performed for each of the four Dengue serotypes (DENV-1, DENV-2, DENV-3, and DENV 4).

Dengue virus sequences were downloaded from NCBI Virus Variation Dengue virus database [11]. Full-length sequences from any host were selected. Next, we extracted only non-coding region 5′UTR and 3′UTR (R1, R2, and R3) using Perl scripts.

After selecting only UTRs, all redundancy between regions was removed. A high conservation of the regions was observed, with a reduction of sequences by at least 72 % (Table 1). For the 3′UTR, the highest conservation was observed for region R3, followed by region R2 and R1.

Table 1. Number of analyzed sequences per Dengue serotype and mean pairwise σ for each non-coding region. Regions R1, R2, and R3 correspond to 3'UTR domains

Serotype	5'UTR		R1		R2		R3	
	Sequences	σ	Sequences	σ	Sequences	σ	Sequences	σ
DENV-1	26	96.99	127	93.59	53	96.78	25	97.96
DENV-2	52	95.16	88	94.55	43	96.72	35	96.93
DENV-3	21	95.39	44	95.88	29	97.62	19	97.31
DENV-4	12	97.31	14	94.14	13	96.09	10	96.32

Furthermore, to verify microRNA target accessibility and to compare serotypes, the secondary structure prediction was performed as presented in Fig. 2. For 5'UTR, we predicted structures highly similar to those previously reported [19]. Also, the structures presented a strong similarity between all four serotypes. The positional entropy is also almost identical, excepts for DENV-3 which is more unstable in the first stem-loop. The 3'UTR presented unexpected conserved structures compared to those previously described [19]. This could be a consequence of limitations of RNAalifold to predict secondary structures for several sequences longer than 100nt with pseudoknots.

Analysing the 3'UTR subdomains, DENV-1, DENV-2, and DENV-3 revealed an extensive region without any conserved structure. Despite the lack of any secondary structure, this region presented a low entropy and well-conserved base pairs. In region R2, the serotypes showed a higher positional entropy which suggests that there are possible alternative structures. In this subdomain, the most similar conserved structures and conserved base pairs are observed in DENV-1 and DENV-3. In region R3, secondary structures were very similar between serotypes, presenting well-conserved base pairs which were more evident in DENV-1 and DENV-2. This could be explained by the importance of this region for viral cyclisation [1].

After structure predictions, miRanda and Perl scripts were used for prediction of microRNAs and selection of targets by structural accessibility, respectively. We could identify 52 microRNAs for DENV-1, 47 for DENV-2, 52 for DENV-3 and 20 for DENV-4. Mostly microRNAs were predicted for 3'UTR, except for hsa-miR-6828-3p and hsa-miR-548g-3p identified in 5'UTR.

Only 5 microRNA targets could be identified in all serotypes (hsa-miR-548g-3p, hsa-miR-6828-3p, hsa-miR-4692, hsa-miR-1914-3p, and hsa-miR-3191-5p). Wen et al. have shown that the microRNA miR-548g-3p suppress DENV multiplication and also affects translation, consequently preventing the expression of viral proteins [27].

Also, some microRNAs were identified only for certain serotypes, as presented in Table 2. Further analysis are necessary to verify the roles of those microRNAs for DENV infections. These results suggested that certain serotypes could be regulated by different microRNAs, which may be used for diagnosis and prog-

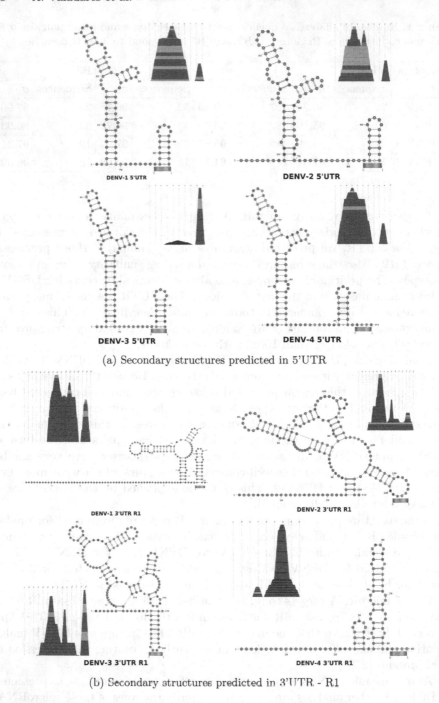

(a) Secondary structures predicted in 5'UTR

(b) Secondary structures predicted in 3'UTR - R1

Fig. 2. Secondary structures predicted for each of the four DENV serotypes in the untranslated regions 5′ (a) and 3′ R1 (b), R2 (c), and R3 (d). In the structure draw, positional entropy is encoded as color hue, ranging from violet and blue for high entropy (low probability pairs) until red for low entropy (high probability pairs).

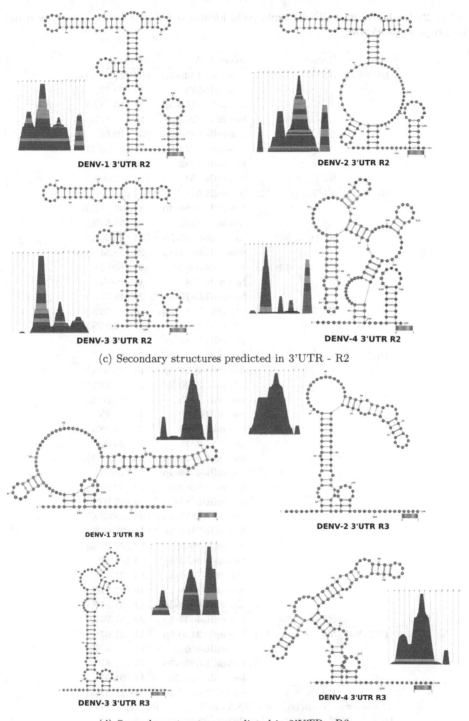

(c) Secondary structures predicted in 3'UTR - R2

(d) Secondary structures predicted in 3'UTR - R3

Fig. 2. (*continued*)

Table 2. Predicted microRNAs exclusively identified for each DENV serotype using the proposed workflow.

Serotype	Region	microRNA	Targets*
DENV-1	R1 (n = 127)	hsa-miR-130b-5p	33 (25.98%)
		hsa-miR-578	10 (7.87%)
		hsa-miR-7158-5p	21 (16.53%)
	R2 (n = 53)	hsa-miR-329-5p	25 (47.17%)
		hsa-miR-6724-5p	53 (100%)
		hsa-miR-6771-5p	18 (33.06%)
		hsa-miR-8082	32 (60.37%)
	R3 (n = 25)	hsa-miR-554	23 (92.00%)
DENV-2	5'UTR (n = 52)	hsa-miR-511-5p	13 (25.00%)
		hsa-miR-6888-3p	47 (90.38%)
		hsa-miR-4312	48 (54.54%)
	R1 (n = 88)	hsa-miR-6505-3p	33 (37.50%)
		hsa-miR-6874-3p	26 (29.54%)
	R2 (n = 43)	hsa-miR-5087	43 (100%)
		hsa-miR-554	43 (100%)
		hsa-miR-6813-3p	20 (46.51%)
	5'UTR (n = 21)	hsa-miR-548ar-3p	13 (61.90%)
		hsa-miR-548az-3p	17 (80.95%)
		hsa-miR-548j-3p	21 (100%)
DENV-3	R1 (n = 44)	hsa-miR-103a-2-5p	39 (88.63%)
		hsa-miR-2682-3p	16 (36.36%)
		hsa-miR-3140-3p	12 (27.27%)
		hsa-miR-3152-5p	11 (25.00%)
		hsa-miR-3660	31 (70.45%)
		hsa-miR-4260	30 (68.18%)
		hsa-miR-4318	11 (25.00%)
		hsa-miR-4474-3p	28 (63.63%)
		hsa-miR-4778-3p	44 (100%)
		hsa-miR-5583-3p	13 (29.54%)
		hsa-miR-6781-3p	18 (40.90%)
		hsa-miR-6875-3p	39 (88.63%)
		hsa-miR-7108-5p	11 (25.00%)
		hsa-miR-25-5p	12 (27.27%)
		hsa-miR-4633-3p	14 (31.81%)
		hsa-miR-7151-3p	14 (31.81%)
		hsa-miR-8068	10 (22.72%)
		hsa-miR-891a-3p	14 (31.81%)
	R2 (n = 29)	hsa-miR-4633-3p	24 (81.75%)
DENV-4	5'UTR (n = 12)	hsa-miR-3180-5p	11 (91.67%)
		hsa-miR-4503	11 (91.67%)
		hsa-miR-520a-5p	11 (91.67%)
		hsa-miR-525-5p	11 (91.67%)
	R2 (n = 14)	hsa-miR-6832-3p	10 (71.42%)

*Number of predicted microRNA binding sites for sequences of each Dengue serotype and non-conding region.
hsa: *Homo sapiens*

nosis of these infections. Besides, it indicates different approaches and solutions depending on the serotype.

4   Conclusion

In this work, we developed a workflow to identify potential targets of micro-RNAs within non-coding regions of Flavivirus genomes. The development of tools to assist the fight against current and emerging arboviral is of great importance, given the increasing incidence and severity of the diseases caused by theses viruses.

Considering the role of non-coding regions in the control of replication and in regulatory functions of Flaviruses, we compared the secondary structures of these regions, followed by the search and identification of microRNA targets in the predicted structures. The final output of our workflow is the selection of the most accessible potential microRNA target binding sites. We applied our workflow in genomes of Dengue virus and compared the secondary structures at the level of serotype. We found some differences mainly in the 3'UTR region between the serotypes. In the 5'UTR region, we could observe different entropies between the serotypes, with DENV-3 being the serotype with the greatest positional instability in spite of having the lowest sequence variability.

In addition to the structural and conservation analysis, we could identify unique targets for each Dengue serotype, such as hsa-miR-130b-5p, hsa-miR-578 and hsa-miR-7158-5p whose targets were identified only in serotype 1 within the R1 region of the 3'UTR. Furthermore, hsa-miR-5087 and hsa-miR-6813-3p were found only in the R2 region of serotype 2 and the microRNA hsa-miR-6832-3p was found in the same region exclusively in serotype 4. The detection of these miRNA targets could be used as a tool for virus classification and diagnosis, facilitating the development of genetic strategies to control these arboviruses.

References

1. Alvarez, D.E., Ezcurra, A.L.D.L., Fucito, S., Gamarnik, A.V.: Role of RNA structures present at the 3 UTR of dengue virus on translation, RNA synthesis, and viral replication. Virology **339**(2), 200–212 (2005)
2. Bartel, P.: MicroRNAs target recognition and regulatory functions. Cell **136**(2), 215–233 (2009)
3. Bhatt, S., et al.: The global distribution and burden of dengue. Nature **496**(7446), 504 (2013)
4. Bidet, K., Garcia-Blanco, M.A.: Flaviviral RNAs: weapons and targets in the war between virus and host. Biochem. J. **462**(2), 215–230 (2014)
5. Darty, K., Denise, A., Ponty, Y.: VARNA: interactive drawing and editing of the RNA secondary structure. Bioinformatics **25**(15), 1974 (2009)
6. Dick, G., Kitchen, S., Haddow, A.: Zika virus (I). isolations and serological specificity. Trans. R. Soc. Trop. Med. Hyg. **46**(5), 509–520 (1952)
7. Diosa-Toro, M., Urcuqui-Inchima, S., Smit, J.M.: Arthropod-borne flaviviruses and RNA interference: seeking new approaches for antiviral therapy. Adv. Virus Res. **85**, 91–111 (2013)

8. Edgar, R.C.: MUSCLE: multiple sequence alignment with high accuracy and high throughput. Nucl. Acids Res. **32**(5), 1792–1797 (2004)
9. Enright, A.J., et al.: Microrna targets in drosophila. Genome Biol. **5**(1), R1 (2003)
10. Gould, E., Solomon, T.: Pathogenic flaviviruses. Lancet **371**(9611), 500–509 (2008)
11. Hatcher, E.L., et al.: Virus variation resource-improved response to emergent viral outbreaks. Nucl. Acids Res. **45**(D1), D482–D490 (2016)
12. Kertesz, M., Iovino, N., Unnerstall, U., Gaul, U., Segal, E.: The role of site accessibility in microRNA target recognition. Nat. Genet. **39**(10), 1278 (2007)
13. Kozomara, A., Griffiths-Jones, S.: miRBase: annotating high confidence micrornas using deep sequencing data. Nucl. Acids Res. **42**(D1), D68–D73 (2013)
14. Kuhn, R.J., et al.: Structure of dengue virus: implications for flavivirus organization, maturation, and fusion. Cell **108**(5), 717–725 (2002)
15. Long, D., et al.: Potent effect of target structure on microRNA function. Nat. Struct. Mol. Biol. **14**(4), 287 (2007)
16. Lorenz, R., et al.: ViennaRNA Package 2.0. Algorithms Mol. Biol. **6**(1), 26 (2011)
17. Lorenz, R., Hofacker, I.L., Stadler, P.F.: RNA folding with hard and soft constraints. Algorithms Mol. Biol. **11**(1), 8 (2016)
18. Witkos, M., Koscianska, T.E., Krzyzosiak, W.J.: Practical aspects of microRNA target prediction. Curr. Mol. Med. **11**(2), 93–109 (2011)
19. Ng, W.C., Soto-Acosta, R., Bradrick, S.S., Garcia-Blanco, M.A., Ooi, E.E.: The 5 and 3 untranslated regions of the flaviviral genome. Viruses **9**(6), 137 (2017)
20. Special Programme for Research and Training in Tropical Diseases, World Health Organization. Department of Control of Neglected Tropical Diseases, World Health Organization. Epidemic and Pandemic Alert and Response: Dengue: Guidelines for Diagnosis, Treatment, Prevention and Control. World Health Organization, Geneva (2009)
21. Rodriguez-Roche, R., Gould, E.A.: Understanding the dengue viruses and progress towards their control. BioMed Res. Int. **2013**, 20 (2013)
22. Schmieder, R., Edwards, R.: Quality control and preprocessing of metagenomic datasets. Bioinformatics **27**(6), 863–864 (2011)
23. Sidahmed, A.M.E., Wilkie, B.: Endogenous antiviral mechanisms of RNA interference: a comparative biology perspective. In: Min, W.P., Ichim, T. (eds.) RNA Interference. Methods in Molecular Biology (Methods and Protocols), vol. 623. Humana Press, New York (2010). https://doi.org/10.1007/978-1-60761-588-0_1
24. Chambers, T.J., Monath, T.P., Maramorosch, K., Shatkin, A.J., Murphy, F.A.: The Flaviviruses: Pathogenesis and Immunity
25. Thurner, C., Witwer, C., Hofacker, I.L., Stadler, P.F.: Conserved RNA secondary structures in flaviviridae genomes. J. Gen. Virol. **85**(5), 1113–1124 (2004)
26. Wang, Y., Zhang, P.: Recent advances in the identification of the host factors involved in dengue virus replication. Virol. Sin. **32**(1), 23–31 (2017)
27. Wen, W., et al.: Cellular microrna-mir-548g-3p modulates the replication of dengue virus. J. Infect. **70**(6), 631–640 (2015)

Parallel Solution Based on Collective Communication Operations for Phylogenetic Bootstrapping in PhyML 3.0

Martha Torres$^{(\boxtimes)}$ and Julio Oliveira da Silva

Núcleo de Biologia Computacional e Gestão de Informações Biotecnológicas
(NBCGIB), Universidade Estadual de Santa Cruz (UESC),
Campus Soane Nazaré de Andrade, Rodovia Jorge Amado, Km 16,
Bairro Salobrinho, CEP, Ilhéus-Bahia 45662-900, Brazil
mxtd2000@yahoo.com.br

Abstract. PhyML is one of the most widely used phylogenetic tree reconstruction programs. Phylogenetic bootstrapping is the classic technique to measure the reliability of a tree that consists of creating disturbances to the original alignment to generate replicates. PhyML 3.0 is an open source program that already includes the bootstrap technique, and this technique is furthermore parallelized in this version using MPI. The parallel solution is based on point-to-point communication operations that are produced within a loop with "number of replicates/number of processors" iterations, causing each processor to construct one phylogenetic tree at a time. The purpose of this work was to modify the parallel version in order to achieve better performance, by firstly replacing the point-to-point communication operations with collective communication operations, and secondly reducing the number of produced messages. The data sets used in the performance evaluation include both synthetic and real data also used by the programs PhyML, RaxML and fatsDNAml. Based on the performance analysis, it was verified that the proposed solution outperforms the original solution, thus proving that collective operations are more efficient than point-to-point operations, and that the grouping of iterations for each processor helps in the overall performance of the application. In addition, it was observed that the proposed solution increases its speedup as the number of bootstrap replicates increases, with a fixed number of processors.

Keywords: Phylogenetic tree reconstruction
Parallel phylogenetic bootstrapping · MPI (Message-Passing Interface)

1 Introduction

Among the programs used for phylogenetic inference, the authors highlight PhyML [1, 2] which is a software that performs the reconstruction process using the Maximum Likelihood method. Recent comparisons of PhyML with other programs, which are also at the top of the rankings for best software for reconstruction of phylogenetic trees (such as RaxML [3] and GARLI [4]), indicate that PhyML is among the fastest and most accurate alternatives. PhyML was as fast and accurate as the RAxML in terms of

© Springer Nature Switzerland AG 2018
R. Alves (Ed.): BSB 2018, LNBI 11228, pp. 133–145, 2018.
https://doi.org/10.1007/978-3-030-01722-4_13

DNA and protein data [1]. Besides, [5] indicates that PhyML has been shown to be faster than RAxML and GARLI when using DNA sequences, and showing the same behavior using protein sequences.

Although the maximum likelihood method is considered as one of the best approaches in the reconstruction of phylogenetic trees, this requires a great computational effort [1, 2]. This high computational cost is mainly due to exhaustive maximum likelihood calculations and to the statistical strategy used for inferred tree certification (in PhyML, bootstrap). PhyML is an open source program. The 3.0 version of PhyML already includes the bootstrap technique to assess the uncertainty of the estimates for the phylogeny, and this technique is furthermore parallelized in this version using the MPI (Message Passing Interface) standard.

In [6] we presented a parallel version, using the shared memory paradigm with OpenMP [7], on the internal maximum likelihood calculations of PhyML 3.0. Furthermore, we compared the performance with the parallel MPI version of the boostrap, giving similar execution times. When performing performance analysis of the hybrid version, the combination of MPI (on bootstrap replicas) and OpenMP (on maximum likelihood calculations), we noticed performance losses that led us to analyze in more detail the original parallel implementation of the boostrap [6], giving rise to the current work where changes are proposed to improve the performance of this version.

The parallel version of the PhyML 3.0 program is based on point-to-point communication operations. These operations are produced "number of bootstrap replicates/number of processors" times, causing each processor to construct one phylogenetic tree at a time. The purpose of this work was to modify this parallel version to achieve better performance, firstly by replacing the point-to-point communication operations with collective communication operations, and secondly by reducing the number of produced messages.

The rest of the article is organized as follows. Section 2 describes the algorithm used by PhyML 3.0 to perform parallel bootstrapping, as well as present the proposed algorithm. Section 3 explains the methodology used in the study. Section 4 subsequently explains the obtained results, and Sect. 5 presents the final conclusions of the present work.

2   Description of the Proposed Parallel Solution

Before describing the proposed solution, the authors explain the parallel bootstrap solution implemented in PhyML 3.0 as shown in Fig. 1.

This algorithm has a main loop (lines 4 to 27), whose iterations correspond to the number of bootstrap samples to be performed divided by the number of processors to be used, i.e. if there are 100 bootstrap replicates and 4 processors, the value of nbRep will be 25, and this loop will be executed 25 times.

For each iteration, the algorithm proceeds first, sending point-to-point messages: processor 0 sends a different vector to each of the other processors, whose positions are randomly shuffled, and which will be used to change input data on each processor.

Once all processors have received their respective vector, each processor proceeds to the calculation of likelihood and to obtaining a tree and its statistics at the same time. This is exposed in Fig. 1 as INSTRUCTION BLOCK FOR CALCULATION OF INDIVIDUAL TREE.

Algorithm 1 original solution code

```
1:  Procedure Bootstrap_MPI (t_tree *tree)
/* bootstrap number/number of processors */
2:    nbRep = tree->mod->bootstrap/Global_numTask
3:    nbElem = Global_numTask;                      //number of processors
4:    for replicate=0:nbRep do
/*Processor 0 sends point-to-point messages to each other processors
with the scrambling vector */
5:      if Global_myRank = 0 then
6:          for i=0:nbElem do
7:              MPI_Ssend (boot_data->wght to i+1)
8:          end for
9:      else
10:       MPI_Recv (boot_data->wght from 0)
11:     end if
12:     INSTRUCTION BLOCK FOR CALCULATION OF INDIVIDUAL TREE
13:     s = Write_Tree(boot_tree,NO);               //store the tree
14:     t= mCalloc(T_MAX_LINE, sizeof(char))
15:     Print_Fp_Out_Lines_MPI(boot_tree,replicate+1,t)//store tree stat.
/* The processors send to the processor 0 the tree and the tree statis-
tics */
16:     if Global_myRank = 0 then
17:       for i=1:nbElem do
18:           MPI_Recv (bootStr from i)
19:           MPI_Recv (bootStr from i)
20:       end for
21:     else
22:       MPI_Ssend (s to 0)
23:       MPI_Ssend (t to 0)
24:     end if
25:     for i=0:2*tree->n_otu-3 do
26:       score_par[i] = tree->a_edges[i]->bip_score
27:     end for
/* This collective operation collects all values of "score_par" from
each processor in vector "score_tot" of processor 0 */
28:     MPI_Reduce(score_par, score_tot, MPI_SUM, 0)
29:   end for
30:   if Global_myRank = 0 then
/* store the final information taking into account all bootstrap repli-
cates */
31:       for i=0:2*tree->n_otu-3 do
32:         tree->a_edges[i]->bip_score = score_tot[i];
33:       end for
34:   end if
35: end procedure
```

Fig. 1. Original solution code

Subsequently, each processor respectively stores its tree information and its statistical data under "t" and "s" (lines 13 and 15). Next, they perform the point-to-point communication again, although this time each processor sends the data corresponding to the calculated tree to processor 0.

Therefore, after performing the *nbRep* iterations, the main loop (lines 4 to 30) ends and thereafter, processor 0 stores the final information taking all replicates of the bootstrap into account, ending the consensus tree. In summary, this solution presents a main loop that has *nbRep* iterations.

This value is directly proportional to the number of replicates and inversely proportional to the number of processors used. The number of messages for each iteration of this loop depends on the number of participating processors. In each iteration, the processors perform the computation of a tree, produce P messages at the beginning and 2P messages of the same size at the end (P: number of processors), in addition to the collective operation MPI_Reduce. The MPI standard creates the knowledge that the collective operations are more efficient than point-to-point operations, since the MPI library has efficient algorithms for these operations.

Thus, the authors propose a modification of the bootstrap algorithm (Fig. 1) to use collective operations, as well as to increase the granularity of the algorithm. Figure 2 describes the proposed solution.

The idea is that processor 0 generates nbRep different vectors (lines 4 to 10), whose positions are randomly shuffled, for each processor and stores them in a vector called "novo_vector_s" (line 7).

Then, the collective operation MPI_Scatter is executed to distribute the *nbRep* vectors for each processor and store them in the "novo_vetor" vector. Next, each processor enters loop, lines 12 to 24, which locally calculates each tree for each iteration by storing their information in vectors "novo_s" and "novo_t" (lines 17 to 20), in addition to saving the partial information for the calculated trees under "score_par" (lines 21 to 23). When the local calculation of the *nbRep* replicates ends (line 24), the collective operation MPI_Reduce is performed to collect the total information for all replicates. In addition, it is necessary to perform the collective operation MPI_Gather twice to store the information of the trees and statistics for all the replicates in processor 0.

The number of messages in this solution depends on the implementation algorithm for the collective operations, and on maximum P messages. Each processor locally calculates the *nbRep* corresponding trees.

In this case, are sent at the start, maximum five messages and at the end, at most *ten* messages, which have larger *nbRep* size than in the Fig. 1. Even so, the number of messages has been reduced because the same number is sent in each iteration in the original algorithm. In addition, the original algorithm is more exposed to problems regarding fine granularity, since each processor calculates one tree at a time and then receives the following message. In the case of the proposed solution, once the data is received it is able to locally calculate *nbRep* trees so that the cost of sending the messages can be masked by the local calculation, i.e. the granularity is higher for this solution.

Algorithm 2 proposed solution code

```
1: Procedure Bootstrap_MPI (t_tree *tree)
/*bootstrap number/number of processors */
2:   nbRep = tree->mod->bootstrap/Global_numTask
3:   nbElem = Global_numTask;                    //number of processors
4:   if Global_myRank = 0 then
//generates "nbRep" random shuffles for each processor
5:       for i=0:nbElem then
6:         for k=0:nbRep then
7:             novo_vetor_s[i*(nbRep)+z+k] = boot_data->wght[z];
8:           end for
9:       end for
10: end if
// collective operation that distributes for each processor its respec-
tive "nbRep" random shuffles
11: MPI_Scatter(novo_vetor_s, novo_vetor, 0)
12: for k=0:nbRep then                             //store the tree
13:    BLOCK OF INSTRUCTIONS FOR THE CALCULATION OF TREES
14:    s = Write_Tree(boot_tree,NO);               //store tree statistics
15:    t= mCalloc(T_MAX_LINE,sizeof(char));
16:    Print_Fp_Out_Lines_MPI(boot_tree, tree->io, k+1, t)//store t. St.
17:    for ind = 0:T_MAX_LINE  then
18:      novo_s[ind + k*T_MAX_LINE] = s[ind]       //store tree statistics
19:      novo_t[ind + k*T_MAX_LINE] = t[ind]       // store the tr
20:    end for
21:    for i=0:2*tree->n_otu-3 then
22:        score_par[i] = tree->a_edges[i]->bip_score;
23:    end for
24: end for
25: MPI_Reduce(score_par, score_tot, MPI_SUM, 0)
/* This collective operation collects all values of "score_par" from
each processor in vector "score_tot" of processor 0 */
26: MPI_Gather(novo_s, bootStr_g, 0)
27: MPI_Gather(novo_t, bootStr_g, 0)
28: if Global_myRank = 0 then
29:    for i=0:2*tree->n_otu-3 then
30:        tree->a_edges[i]->bip_score = score_tot[i]
31:    end for
32: end if
33: end procedure
```

Fig. 2. Proposed solution code.

3 Materials and Methods

The developed implementation was tested on the high-performance computer CACAU (Center for Data Storage and Advanced Computing at the State University of Santa Cruz). CACAU consists of 20 nodes, totaling 1.0 TeraFLOP. Each node has 2 Intel (R) Xeon (R) E5430, 2.66 GHz, QuadCore, 16 GB RAM memory. The nodes are interconnected across the Infiniband network. The authors have used the GNU/Linux 2.6.32-642 operating system, the gcc 5.4.0 compiler and the OpenMPI 2.1.0 library.

The methodology used to evaluate the proposed solution is the one most authors use for this purpose [1, 10–13]. This methodology mainly consists of using a set of data that most closely represents the real data. In our case, it was chosen a small dataset representative of those available for performance evaluation. The data sets used in the performance evaluation are divided into two groups: the first group refers to data sets whose sequential execution time without bootstrap is shorter than one minute, which are called small-size data sets, and the other group refers to data sets whose sequential execution time without bootstrap is between 2 and 14 min, called medium-size data sets.

In total, 18 small-size data sets were used of which 9 (data22.phy, dada24.phy, data36.phy, data42.phy, data49.phy, data54.phy, data74.phy and data84.phy) are simulated data available from[1]; this data sets includes 40 sequences and 500 sites, and they have been generated by Seq-Gen [8] along random trees, using the GTR model, with parameters estimated from HIV data [9]: nucleotide frequencies fA = 0.40, fC = 0.20, fG = 0.22, fT = 0.18, four rate categories of gamma shape parameter 0.969, and rates of nucleotide changes r(AC) = 1.72, r(AG) = 5.03, r(AT) = 0.84, r(CG) = 0.91, r(CT) = 7.70, r(GT) = 1 [10].

The other 9 data sets (protein_M1989, protein_M1381, protein_M1382, protein_M1384, protein_M1385, protein_M1889, protein_M1882, protein_M2638, protein_M2640 and protein_M2641) are real data sets consisting of actual protein alignments and available at[2]. These data sets were used in [1], extracted from Treebase [11]. The selection criteria used to choose the alignments was between 5 and 200 sequences shorter than 2,000 sites, in addition to being part of the 50 most recent protein alignments registered under Treebase [1]. The authors have also used 4 medium-size data sets, two of which (Nucleic_M2792 and proteic_M2477 available at[3]) are real data selected from Treebase and have been used in [1]. The other two are 101_SC and 150_SC which have respectively 101 and 150 sequences of several fungi, and were used in [12, 13].

The input parameters used in Phyml 3.0 to perform the performance analysis were -b 100 -s BEST -o tlr -c 4. The model used for the nucleotide was -m GTR and -m WAG for the protein. In addition, the execution time provided by the Phyml 3.0 program was used in the present study.

4   Results

The performance analysis has been divided into two parts. The first part uses small-size data sets, and the second part uses medium-size data sets.

4.1   Small-Size Data Sets

In order to verify the correct implementation of the proposal and taking advantage of the fact that the simulated data contain the true trees, a comparison was made between the true trees and those provided by the original sequential solution, and the proposed parallel solution.

---

[1] http://www.atgc-montpellier.fr/phyml/benchmarks/data/simu/.

[2] http://www.atgc-montpellier.fr/phyml/benchmarks/index.php?ben=md.

[3] http://www.atgc-montpellier.fr/phyml/benchmarks/.

The comparison was made using the Ktreedist program [14], which taken both topology and branch length information of a phylogenetic tree into account. This program computes a K-score that measures overall differences in the relative branch length and topology of two phylogenetic trees after scaling one of the trees to have a global divergence which is as similar as possible to the other tree. High K-scores indicate a poor match between the estimated tree and the reference tree. Lower K scores from Ktreedist indicate that two trees are more similar in terms of differences of the relative branch length and topology.

Table 1 shows the K-score values, in this case the true tree available at[4] was used as the reference tree comparing it with the result of the original sequential program (Original seq) and the proposed solution using 2 and 32 processors (2P and 32P).

Table 1. K-score values of original sequential solution (Original seq) and the proposed solution for 2 (2P) and 32 processors (32P).

Data	K-score		
	Original seq	Proposed(2P)	Proposed(32P)
data22.phy	0.03858	0.03855	0.03865
data24.phy	0.04133	0.04120	0.04132
data26.phy	0.06541	0.06524	0.06565
data36.phy	0.04813	0.04804	0.04814
data42.phy	0.05305	0.05303	0.05305
data49.phy	0.03984	0.03977	0.03979
data54.phy	0.06784	0.06822	0.06803
data74.phy	0.05801	0.05822	0.05818
data84.phy	0.02648	0.02654	0.02652

The results indicate that the proposed solution provides equivalent results to the original version and this result is practically unaffected when using a different number of processors, the difference between the results ranging from 0.00002 to 0.0006.

Another way to verify the correct implementation of the proposed solution is to compare the final log likelihood values to real data. Therefore, Table 2 summarizes the final log likelihood values of real small-size data sets for 2 and 32 processors, comparing the proposed solution with the original solution. Due to the nature of the Phyml program, each execution of the program provides a different final tree, even though it may be observed that the likelihood values are very close to one another. The maximum difference was in the order of 0.02581. Therefore, our implementation presented the same behavior that original solution.

Table 3 lists the execution times in seconds for the original (O) and proposed (P) version for both 8 and 16 processors, and the run time improvement (I, in %) for the proposed solution over the original.

---

[4] http://www.atgc-montpellier.fr/phyml/benchmarks/index.php?ben=sm.

Table 2. The final log likelihood values for 2 and 32 processors, comparing the proposed and original solution to real data

Data	2P		32P	
	Proposed	Original	Proposed	Original
M1381	−12077.99995	−12078.0031	−12077.99932	−12078.00295
M1382	−12335.18897	−12335.18706	−12335.18880	−12335.19031
M1384	−2882.79360	−2882.79319	−2882.79329	−2882.79321
M1385	−5609.46830	−5609.47086	−5609.46969	−5609.46957
M1882	−6191.98897	−6191.98891	−6191.98863	−6191.98891
M1989	−3179.91146	−3179.90668	−3179.91570	−3179.91490
M2638	−1690.35010	−1690.33670	−1690.33655	−1690.33346
M2640	−2510.78407	−2510.78343	−2510.78139	−2510.78273
M2641	−5153.58044	−5153.57942	−5153.58347	−5153.58195

Table 3. Execution times in seconds for the proposed (P) and original (O) solutions for 8 (8P) and 16 (16P) processors, and improvement (I) percentage of the proposed solution.

Data	8P			16P		
	O	P	I	O	P	I
Data22	297	269	9.43	317	172	45.74
Data24	251	215	14.34	246	129	47.56
Data26	284	268	5.63	285	158	44.56
Data36	278	241	13.31	277	151	45.49
Data42	240	223	7.08	259	140	45.95
Data49	230	220	4.35	244	133	45.49
Data54	253	229	9.49	266	148	44.36
Data74	218	197	9.63	229	127	44.54
Data84	250	214	14.40	249	127	49.00
M1381	303	273	9.90	294	176	40.14
M1382	346	274	20.81	327	170	48.01
M1384	231	201	12.99	231	132	42.86
M1385	288	262	9.03	289	156	46.02
M1882	184	176	4.35	199	109	45.23
M1989	360	334	7.22	393	185	52.93
M2638	383	331	13.58	389	200	48.59
M2640	420	383	8.81	440	240	45.45
M2641	397	353	11.08	385	226	41.30

For 8 and 16 processors, the proposed version always outperformed the original version. In the case of 8 processors, the proposed version showed average gains of 10.30%. For 16 processors, the average improvement was 45.73%. It can be noticed that the original version does not scale, i.e. the execution time using 8 processors was shorter than the one using 16. In contrast, the proposed version continues to scale with 16 processors, and a significant performance gain was therefore obtained. It is important to point out that in the original solution, the overhead increases with the increase in the number of processors, thus impairing its performance.

Table 4 lists the execution times in seconds for the original (O) and proposed (P) version for both 32 and 64 processors, and the run time improvement (I, in %) for the proposed solution over the original. Again, the original version does not scale while the proposed version continues to scale.

Table 4. Execution times in seconds for the proposed (P) and original (O) solutions for 32 (16P) and 64 (64P) processors, and improvement (I) percentage of the proposed solution.

Data	32P			64P		
	O	P	I	O	P	I
Data22	486	114	76.54	974	81	91.68
Data24	411	90	78.10	805	64	92.05
Data26	477	108	77.36	957	80	91.64
Data36	434	102	76.50	907	69	92.39
Data42	416	99	76.20	832	70	91.59
Data49	399	88	77.94	794	61	92.32
Data54	434	104	76.04	834	66	92.09
Data74	376	86	77.13	752	61	91.89
Data84	375	85	77.33	778	59	92.42
M1381	493	124	74.85	932	80	91.42
M1382	505	119	76.44	1013	83	91.81
M1384	402	93	76.87	763	59	92.27
M1385	477	116	75.68	909	86	90.54
M1882	308	77	75.00	620	50	91.94
M1989	598	130	78.26	1154	90	92.20
M2638	580	131	77.41	1204	92	92.36
M2640	710	154	78.31	1449	108	92.55
M2641	653	161	75.34	1302	108	91.71

Next, a study was carried out on the behavior of the proposed solution with respect to its scalability and efficiency. Table 5 shows speedup (S) described in formula (1), and efficiency (E), described in formula (2), for 32 and 64 processors.

$$\text{Speedup}_P = \text{Sequential Execution Time}/\text{Parallel Execution Time}_P \qquad (1)$$

$$\text{Efficiency}_P = \text{Speedup}_P/P \qquad (2)$$

It shows an important feature of the proposed solution, which is that the solution continues to reduce the execution time even using 64 processors. This is reflected in the increase in speedup which still presents gain with 64 processors, with respect to the speedup of 32 processors. The efficiency is decreasing because the speedup values are increasingly far from the ideal value (linear speedup); this means that by increasing the number of processors, the overhead generated by the passage of messages in collective operations increases the total execution time, and each time the local execution time is reduced. The average speedup for 32 and 64 processors was respectively 15.08 and 21.9. The average efficiency for 32 and 64 processors was respectively 41.13% and 34.22%.

4.2   Medium-Size Data Sets

The results obtained for medium-size data sets are presented as follows. Table 6 shows the execution times in seconds for the original (O) and proposed (P) version for both 16 and 24 processors, and the run time improvement (I, in %) for the proposed solution over the original.

Table 5. Speedup (S) and efficiency (E) for 32 and 64 processors for medium-size data sets.

Data	S32	S64	E32	E64
Data22	15.69	22.09	49.04	34.51
Data24	15.01	21.11	46.91	32.98
Data26	15.45	20.86	48.29	32.60
Data36	15.22	22.49	47.55	35.14
Data42	14.71	20.80	45.96	32.50
Data49	15.75	22.72	49.22	35.50
Data54	14.29	22.52	44.65	35.18
Data74	15.37	21.67	48.04	33.86
Data84	16.46	23.71	51.43	37.05
M1381	13.94	21.60	43.55	33.75
M1382	15.18	21.77	47.45	34.02
M1384	14.78	23.31	46.20	36.41
M1385	14.29	19.28	44.67	30.12
M1882	13.60	20.94	42.49	32.72
M1989	15.91	22.98	49.71	35.90
M2638	16.66	23.99	52.08	37.48
M2640	16.28	22.79	50.87	35.61
M2641	12.88	19.57	40.26	30.57

This table shows that the proposed solution improves the execution times by an average of 45.95. In the case of 24 processors, the original solution has a longer execution time than using 16, i.e. it does not scale anymore, while the proposed solution continues to scale.

Table 6. Execution times in seconds for the proposed (P) and original (O) solutions for 16 (16P) and 24 (24P) processors, and improvement (I) percentage of the proposed solution.

Data	16P			24P		
	O	P	I	O	P	I
proteic_M2477	2467	1469	40.45	3313	1123	66.10
nucleic_M2792	9682	5241	45.87	13150	4560	65.32
101_SC	12107	5824	51.89	13598	6309	53.60
150_SC	14324	7775	45.72	18462	6372	65.49

Table 7 shows the execution times in seconds for the original (O) and proposed (P) version for 32, 64 and 72 processors, and the run time improvement (I, in %) for the proposed solution over the original. This table shows that the proposed solution also scales for medium-size data sets.

Table 7. Execution times in seconds for the proposed (P) and original (O) solutions for 32 (16P), 64 (24P) and 72 (72P) processors, and improvement (I) percentage of the proposed solution.

Data	32P			64P			72P		
	O	P	I	O	P	I	O	P	I
proteic_M2477	4264	1015	76.20	8561	658	92.31	9553	642	93.3
nucleic_M2792	16918	3458	79.56	32242	2865	91.11	35709	2549	92.86

Next, a study was carried out on the behavior of the proposed solution with respect to its scalability and efficiency. Table 8 shows speedup (S) described in formula (1), and efficiency (E), described in formula (2), for 32 and 64 processors. The proposed solution continues to scale when using 64 processors, and the speedup value continues to rise although efficiency is decreasing as it is getting further away from the ideal speedup value. The average speedup was 14.76 and 21.21 for 32 and 64 processors respectively, and the average efficiency was 46.15% and 36.14% for 32 and 64 processors respectively.

Table 8. Speedup (S) and efficiency (E) for 32 and 64 processors for medium-size data sets.

Data	S32	S64	E32	E64
proteic_M2477	14.24	21.97	44.51	34.33
nucleic_M2792	16.47	19.88	51.47	31.06
101_SC	12.61	19.31	39.41	30.17
150_SC	15.71	23.69	49.11	36.99

In order to show how the proposed solution behaves with increasing number of bootstrap replicates, Table 9 shows the speedup for 32 processors for bootstrap values of 100, 300 and 1000.

Table 9. Speedup of 32 processors (S32) using 100 (100B), 300 (300B) and 1000 (1000B) bootstrap of medium-size data sets

S32	100B	300B	1000B
proteic_M2477	14.24	22.33	27.63
nucleic_M2792	16.47	22.23	28.92

Table 9 enables the conclusion that the speedup improved when increasing the number of bootstrap replicates, i.e. the proposed solution scales when increasing the number of bootstraps. This means that, as the size of the local task significantly increases, the more efficient the solution becomes.

5   Conclusions

The proposed solution always outperforms the original solution, showing that collective operations are more efficient than point-to-point operations, and that the grouping of iterations for each processor helps in the overall performance of the application. The proposed solution always scaled in the experiments, and up to 80 processors were used always resulting in less time.

Based on the analysis, it was verified that the proposed solution increases its speedup, which translates into increased efficiency as the number of bootstraps increases with number of fixed processors. In other words, the solution becomes more efficient as it increases the amount of local iterations by diluting the communication overhead for collective operations.

Based on the obtained results, it is possible to conclude that as the number of processors increases, the communication overhead of point-to-point operations and the increase in the number of messages considerably affects the original solution. Moreover, for fine-granularity data sets this solution only scaled up to 8 processors, and with coarser granularity it scaled up to 16 processors. Additionally, the proposed solution presented similar results for speedup and efficiency for short and medium-size data sets, that is to say the influence of the individual execution time of each iteration was smaller in this type of solution.

For purposes of future analysis, one can consider large-size data sets that will certainly positively affect the performance of the proposed solution because the increase in execution time for one iteration represents increasing the local granularity masking the communication overheads. Future work is also expected to explore collective operations of the MPI3 standard to increase the efficiency of results.

Acknowledgments. This study was supported by The State University of Santa Cruz (UESC).

References

1. Guindon, S., et al.: New algorithms and methods to estimate maximum-likelihood phylogenies: assessing the performance of PhyML 3.0. Syst. Biol. **59**(3), 307–321 (2010). https://doi.org/10.1093/sysbio/syq010
2. Guindon, S., Gascuel, O.: A simple, fast, and accurate algorithm to estimate large phylogenies by maximum likelihood. Syst. Biol. **52**(5), 696–704 (2003). https://doi.org/10.1080/10635150390235520
3. Stamatakis, A.: RAxML-VI-HPC: maximum likelihood-based phylogenetic analyses with thousands of taxa and mixed models. Bioinformatics **22**(21), 2688–2690 (2006). https://doi.org/10.1093/bioinformatics/btl446
4. Zwickl, D.J.: Genetic algorithm approaches for the phylogenetic analysis of large biological sequence datasets under the maximum likelihood criterion. Thesis (Doctor of Philosophy), Faculty of the Graduate School, University of Texas at Austin, Austin, Texas, 115 f (2006)
5. Criscuolo, A.: morePhyML: improving the phylogenetic tree space exploration with PhyML 3. Mol. Phylogenet. Evol. **61**(3), 944–948 (2011). https://doi.org/10.1016/j.ympev.2011.08.029
6. Silva, J.O., Orellana, E., Torres, M.: Development of a parallel version of PhyML 3.0 using shared memory. IEEE Latin Am. Trans. **15**(5), 959–967 (2017). https://doi.org/10.1109/TLA.2017.7912593
7. Chandra, R., Dagum, L., Kohr, D., Maydan, D., McDonald, J., Menon, R.: Parallel Programing in OpenMP. Morgan Kaufmann, San Francisco (2001)
8. Rambaut, A., Grassly, N.C.: Seq-Gen: an application for the Monte Carlo simulation of DNA sequence evolution along phylogenetic trees. Comput. Appl. Biosci. **13**, 235–238 (1997). https://doi.org/10.1093/bioinformatics/13.3.235
9. Posada, D., Crandall, K.A.: Selecting the best-fit model of nucleotide substitution. Syst. Biol. **50**(4), 580–601 (2001)
10. Anisimova, M., Gascuel, O.: Approximate likelihood-ratio test for branches: a fast, accurate, and powerful alternative. Syst. Biol. **55**, 539–552 (2006). https://doi.org/10.1080/10635150600755453
11. Sanderson, M.J., Donoghue, M.J., Piel, W., Eriksson, T.: TreeBASE: a prototype database of phylogenetic analyses and an interactive tool for browsing the phylogeny of life. Am. J. Bot. **81**, 183 (1994)
12. Stamatakis, T., Ludwig, Meier H.: RAxML-III: a fast program for maximum likelihood-based inference of large phylogenetic trees. Bioinformatics **21**(4), 456–463 (2005). https://doi.org/10.1093/bioinformatics/bti191
13. Olsen, G., Matsuda, H., Hagstrom, R., Overbeek, R.: fastDNAmL: a tool for construction of phylogenetic trees of DNA sequences using maximum likelihood. Comput. Appl. Biosci. **10**, 41–48 (1994). https://doi.org/10.1093/bioinformatics/10.1.41
14. Soria-Carrasco, V., Talavera, G., Igea, J., Castresana, J.: The K tree score: quantification of differences in the relative branch length and topology of phylogenetic trees. Bioinformatics **23**, 2954–2956 (2007). https://doi.org/10.10193/bioinformatics/btm466

Author Index

Printed in the United States
By Bookmasters